MAN IN THE COSMOS

An Inquiry into the Ideas of G.I. Gurdjieff
from a Scientific Perspective

MAN IN THE COSMOS

An Inquiry into the Ideas of G.I. Gurdjieff from a Scientific Perspective

BY CHRISTIAN WERTENBAKER

 Codhill Press

Codhill Press books
are published for David Appelbaum

First Edition
Printed in the United States of America
Copyright © 2012 by Christian Wertenbaker

ISBN 1-930337-69-8

Cover and interior text design by Alicia Fox.

Library of Congress Cataloging-in-Publication Data
Wertenbaker, Christian, 1943–

Man in the cosmos : an inquiry into the ideas of G.I. Gurdjieff from a scientific perspective / by Christian Wertenbaker. — 1st ed.

p. cm.

ISBN 1-930337-69-8

1. Fourth Way (Occultism) 2. Gurdjieff, Georges Ivanovitch, 1872–1949. I. Title.

BP605.G94W47 2012

197--dc23

2012040140

For Dolphi, Carlo, and Elena

ACKNOWLEDGEMENTS

The author gratefully acknowledges the following rights holders
for giving their permission to quote at some length:

Triangle Editions, Inc., from G.I. Gurdjieff,
Beelzebub's Tales to his Grandson, All and Everything, First Series.

Tatiana Nagro, from P.D. Ouspensky,
In Search of the Miraculous. Fragments of an Unknown Teaching.

TABLE OF CONTENTS

INTRODUCTION

This book is a collection of essays that try to relate two distinct areas of human knowledge: the mystical cosmology of G.I. Gurdjieff, based, according to him, on ancient wisdom, and the discoveries and theories of modern science. I first encountered Gurdjieff's ideas at the age of twenty, and early on developed three basic convictions: First, that, on reading the personal accounts of people who had interacted with Gurdjieff, it was clear that he possessed a degree of awareness, attention, perception, knowledge, and ability to act that put him on another level compared to ordinary people; because of that, and the lucidity of his descriptions of the human condition, it was necessary to give credence to his more obscure and controversial ideas about the nature of the universe, of man, of the soul, and of their relationships. Secondly I was convinced that the method of modern science is a generally valid and honest way to arrive at truths about the world, with one caveat: because science deliberately tries to be objective and to remove the subjectivity of the observer from its deliberations, its findings may only apply to the external world, leaving the inner world of conscious beings to another realm of inquiry. Third, that there is nevertheless only one world, and so all truths about it must be compatible and related.

These principles have guided these essays, many of which have been published in Parabola magazine over the last fifteen years. They represent a personal quest for a more comprehensive truth about the nature of the world. To pursue my aim of reconciling Gurdjieff's ideas and modern science, I studied the sciences broadly in college, then went on to medical school and to postgraduate training in neurophysiology, neurology, neuro-ophthalmology and ophthalmology. I also became a member of the Gurdjieff Foundation, devoted to exploring and pursuing Gurdjieff's ideas and aims.

I have not arrived at definite conclusions, and still do not know for sure whether many of Gurdjieff's ideas are true. These essays are essays in the original sense of the word: attempts to understand. Because each essay was initially separate and self-sufficient, there is a fair amount of repetition in this book, but this may not be a bad thing: a particular concept can be approached from many angles, and with difficult ideas this can aid understanding. Because Parabola magazine has a theme for each issue, each essay that was published there takes its viewing angle from the theme of the issue it was in. At the end of this introduction is a list of the essays that were published in Parabola, the title of the essay in Parabola, and the theme of the issue it was published in.

The essays are here arranged into categories, but they do not make an entirely logical sequence, and the order of reading them can be arbitrary. I would suggest reading them one or two at a time. Most of the essays do not presume a familiarity on the part of the reader with Gurdjieff's ideas, but "Some Thoughts on the Enneagram" does.

I am grateful to the many people, too numerous to list, with whom I have exchanged ideas, and to the editors of Parabola I have worked with, especially David Appelbaum, who is also the editor of this collection, and Jeff Zaleski. Thanks to Elena and Dolphi Wertenbaker, Priscilla Smith, Alicia Fox, Miriam Faugno, and Dianne Edwards for help putting this together.

Perhaps others can build on these ideas, or refute them.

Essays published in Parabola

Nature's Patterns: Published as *Nature's Patterns—Hidden lines of emergence*, in Nature, vol. 24, No. 1, 1999.

Pythagoras in 1999: Published as *A New Science of the Unknown—An interview with Pythagoras*, in Number and Symbol, vol. 24, No. 3, 1999.

The Light of the Beholder: Published as *The Eye of the Beholder—Divinity and consciousness in tradition*, in Light, vol.26, No. 2, 2001.

One and One Make One: Published as *One and One Make One—The puzzle of looking inward and outward*, in Riddle and Mystery, vol. 25, No. 2, 2000.

Shadows of the Real World: Published as *Toward a Vision of the Whole—Understanding the eyes' work*, in The Shadow, vol. 22, No. 2, 1997.

Imagination: Published as *Imagination—A scientist explores the many facets of imagination*, in Imagination, vol. 34, No. 1, 2009.

Awakening the Emotions: Published as *Awakening the Emotions—Transforming the emotional life*, in Fear, vol. 23, No. 3, 1998.

The Ego and the I: Published as *Joining the Sparks with the Fire—The cosmos shorn of duality*, in The Ego and the "I"—Which One is Real? vol. 27, No. 1, 2002.

The Home of the Self: Published as *The Home of the Self—The many cosmoses we live in, and humanity's special place within them*, in Home, vol. 31, No. 4, 2006.

Holy Earth: Published as *Is the Earth Conscious?—A neuroscientist explores the mind of planet Earth*, in Holy Earth, vol. 32, No. 3, 2007.

Laws, Miracles and Science: Published as *Laws, Miracles and Science—A multi-leveled cosmos*, in Miracles, vol. 22, No. 4, 1997.

The Materiality of the Soul: Published as *The Materiality of the Soul*, in Science & Spirit, vol. 37, No. 4, 2012.

The Fullness of the Void: Published as *The Fullness of the Void—A scientific approach to the mystery of consciousness*, in Thinking, vol. 31, No. 3, 2006.

The Cosmic Necessity of Suffering: Published as *The Cosmic Necessity of Suffering—Suffering in the lives of humanity, the universe, and God*, in Suffering, vol. 36, No. 1, 2011.

The Cosmic Metabolism of Form: Published as *The Cosmic Metabolism of Form—A meditation on a "great cosmic ecology of consciousness"*, in Seeing, vol. 36, No. 3, 2011.

Vibrations: Translated into Arabic and published in an Arabic website devoted to Gurdjieff's ideas.

1

MATHEMATICS, THE SCIENCE OF PATTERNS

Nature's Patterns

Tyger! Tyger! burning bright
In the forests of the night,
What immortal hand or eye
Could frame thy fearful symmetry?

—William Blake

Consider the work of God; who can make straight what he has made
crooked? In the days of prosperity be joyful, and in the days of adversity
consider: God has made the one as well as the other, so that mortals
may not find out anything that will come after them.

(Ecclesiastes 7:13-14)

These quotations describe two apparently contradictory aspects of the natural world that we see around us: on the one hand, evidence of an supremely fine-tuned orderliness in nature's vast cornucopia of phenomena, and on the other, chaos, conflict, unpredictability, and catastrophe—"there are no straight lines in nature"[1]. Any number of dichotomies can be pointed to: eclipses can be predicted centuries in advance, but the weather cannot be forecast reliably for even a few days; the beautiful fivefold symmetry of a starfish contrasts with the total irregularity of a coastline; the simple arc of a rainbow meets the indescribable shape of a cloud. Snowflakes exemplify both aspects: no two are alike, yet they all have hexagonal symmetry.

For many centuries mathematics—the science of patterns—concerned itself primarily with numbers and with regular shapes: the straight lines, triangles, circles, and ellipses familiar to generations of schoolchildren since the days of Euclid. Difficulties arose straightaway: the Greeks were very disturbed by the fact that the ratio of a circle's circumference to its diameter was an irrational number, *pi,* a number that could not be represented by either a digit or a ratio of two digits, although $^{22}/_7$ came close. Nor could the ratio of a square's diagonal to its side, or any number of other geometric entities. be so represented. The decimal digits of pi, 3.141592653. . . . , going on endlessly, continue to resist any attempt to find a pattern in them, despite having been calculated to billions of digits.[2] Yet, beginning in the sixteenth century, pi was found to be precisely expressible as an orderly infinite series, one version of which is: pi = 4 $(1 - \frac{1}{3} + \frac{1}{5} - \frac{1}{7} + \frac{1}{9} - \frac{1}{11}$ . . . etc). This exemplifies how mathematics has developed, progressively finding patterns hidden in apparent randomness, at new levels of abstraction.

Modern science attempts to discover mathematical patterns in natural phenomena. It is often felt to have begun in earnest with Galileo, Kepler, and Newton. Kepler discovered that the planets move around the sun in ellipses, rather than in the complicated patterns of circles upon circles—epicycles— which had been imposed on their observed motions since the days of Ptolemy. This then led Newton to find the simple underlying laws of motion and gravitation which many of us learn in school, the knowledge of which makes possible artificial satellites and the like. The subsequent triumphs of this approach to making sense of phenomena—the application of mathematics to wave motion, electricity and magnetism, chemical reactions, atomic structure, and so forth—and the technological spinoffs, which surround us every day, are well known. At the same time, many aspects of the natural world failed to succumb to this approach: the shapes of trees and clouds, the complexity of living organisms, the haphazard bounty of ecosystems. Some questions about these things were not even asked, many complex phenomena being considered to be the result of incalculable random forces. (One of the unfortunate aspects of a specialized scientific education is that one quickly learns which questions are acceptable within the framework and conventions of the discipline; for instance, in medical school, one quickly learns not to ask why people laugh.)

Partly because mathematical science seemed either to have little relevance to everyday concerns, or provided pseudo-answers which appeared both simplistic and inhuman, early in the twentieth century the humanities and the sciences appeared to be on the verge of divorce. Meanwhile, mathematicians went on to discover more and more abstract mathematical phenomena, most of which seemed not to have anything to do with the real world at all, so mathematicians and physicists were also barely on speaking terms. But in the last several decades, a new field of mathematical exploration has appeared which applies to many of these previously unapproachable aspects of nature. This is a many-faceted realm of inquiry, largely made possible by the digital computer, parts of which are known as chaos theory, complexity theory, catastrophe theory, nonlinear dynamics, and fractal geometry.

It is worthwhile here to pause briefly to consider the age-old epistemological question: What is the relationship of mathematics to reality, and of mathematical knowledge to real understanding? Pythagoras and Plato considered mathematical reality to be the ultimate reality, and the uncanny tendency of even the most obscure and abstract mathematical phenomena to find application, sooner or later, to the world of nature has astonished scientists and philosophers for millennia. Paul Valery, the poet, said: "The universe is built on a plan the profound symmetry of which is somehow present in the inner structure of our intellect." [3] At the same time, the twentieth century produced Godel's theorem, which shows that no sufficiently inclusive mathematical system is complete in itself. One version reads: "For every consistent formalization of arithmetic, there exist arithmetic truths unprovable within that formal system."[4] Such considerations have led thinkers like Penrose to conclude that humanlike consciousness and understanding cannot in principle be developed in a computer, no matter how complex.[5] Certainly the direct experience of shape by touch or of color by sight would seem to be irreducible, even though it might correlate precisely with mathematical descriptions of shape or of the wavelengths of light. I would prefer to think of mathematical (logical) perception as one—fully valid, but only one—of several dimensions of perception of which we are capable. As John von Neumann says:

> "The sciences do not try to explain, they hardly even try to interpret, they mainly make models. By a model is meant a mathematical construct which, with the addition of certain verbal interpretations, describes observed phenomena. The justification of such a mathematical construct is solely and precisely that it is expected to work."[6]

Nevertheless, "mathematics is the science of patterns, and nature exploits just about every pattern that there is."[7]

Perhaps even more astonishing than the fact that mathematical patterns correspond to patterns found in nature are the mathematical patterns themselves. Why should exactly six equilateral triangles fit into a circle, with sides the same length as the radius of the circle?

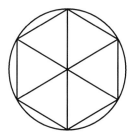

And why should equilateral triangles, hexagons, and squares be the only regular figures that tile the plane without gaps? All of these patterns are seen abundantly in nature—in snowflakes and honeycombs, for instance—nature making use of these regularities to build its structures. The regular solids in three dimensions, the five platonic solids, also show up in nature, and the pentagonal symmetry seen in certain plants and animals may also be related to the fact that two of the five platonic solids, the dodecahedron and the icosahedron, exhibit pentagonal symmetries.

(Top) Common milkweed blossom
(Right) Computer-generated views of
an icosahedral herpes virus

Why does an object thrown up and sideways into the air describe a parabola? Or the planets move in ellipses? Both these curves are conic sections, formed by slicing a cone in various ways. It was Newton's genius to show that these paths follow from the simple laws of motion and gravitation that he postulated; to do so he had to invent a new branch of mathematics, the calculus.

Using the word "why" in the previous paragraphs seems to fall into the trap of looking for explanations which science cannot provide; recall von Neuman's statement above. The real "why" of these patterns remains fundamentally mysterious and evokes a feeling of wonder, even religious awe; many of the great scientists have had such feelings.

Why does a pebble thrown into a still pond produce a pattern of concentric circular ripples? Mathematics, being the science of patterns, is therefore the science of symmetries. Symmetry, whether in the ripples of a pond or in a flower, or in man-made designs, resonates deeply with our sense of the beautiful, as Blake and Valery attest. But it is really mixtures of symmetry and irregularity, or of different kinds of symmetries superimposed, which most strike us, and which carry the most information. The mathematics of symmetry has become more and more sophisticated, and has assumed a greater and greater role in the scientific modeling of nature. A symmetry is an invariant quality in the face of a transformation. The symmetry of a hexagon, for instance, refers to the fact that its shape does not change if it is rotated in multiples of sixty degrees (a sixth of a full turn). It also does not change if it is reflected about any axis passing through opposite corners. A hexagon is then less symmetrical than a circle, which does not change its shape if rotated any amount whatsoever, or reflected about any diameter.

The most symmetrical object of all, then, is endless empty space, which does not change under any rotation, displacement, or reflection. As pointed out by Prirogine,[8] a hypothetical tiny observer in such a space would see no change with any movement, and thus would have no way to perceive the idea of space. If nothing happened in this space, the observer would have no way to perceive the idea of time. The existence of our world depends on what mathematicians call broken symmetries. If everything were uniform it could carry no information, and likewise if everything were completely random and irregular. So our world depends on a mixture of the two. When a pebble falls through the smooth surface of a lake, it breaks the symmetry of the surface by establishing a point which is different from all the others; the ripples have a circular symmetry around this point. Symmetry has been broken, but as much as possible is preserved in the face of the disturbance.

Current speculations in physics postulate that the four fundamental forces of nature—gravity, electromagnetism, and the strong and weak nuclear forces—arose from a broken symmetry, that at the high temperatures of the early universe which existed just after the "big bang," the four forces were one. As water cools and becomes ice, symmetry is broken but other symmetry appears; similarly as the universe cooled, the fundamental forces separated out from one primordial force. This concept depends on advanced mathematical representations of these forces, but the same basic concept is present in the Bible:

> In the beginning God created the heaven and the earth. And the earth was without form, and void; and darkness was upon the face of the deep. And the Spirit of God moved upon the face of the waters. And God said, Let there be light; and there was light. And God saw the light, that it was good: and God divided the light from the darkness. And God called the light Day, and the darkness he called Night. And the evening and the morning were the first day. And God said, let there be a firmament in the midst of the waters, and let it divide the waters from the waters. And God made the firmament, and divided the waters which were under the firmament from the waters which were above the firmament: and it was so. And God called the firmament

Heaven. And the evening and the morning were the second day. And God said, let the waters under the heaven be gathered together unto one place, and let the dry land appear: and it was so. (Genesis 1:1-9)

A specific version of broken symmetry appears in G.I. Gurdjieff's description of the Creation. In order to create our universe, God altered the fundamental laws of nature to make them less symmetric:

> Our Common Father Uni-being Endlessness, having decided to change the principle of the maintenance of existence of that still unique cosmic concentration and sole place of His Most Glorious Being, first altered the functioning itself of these two primordial sacred laws, and He made the greater change in the law of the sacred Heptaparaparshinokh.
>
> This change in the functioning of the sacred Heptaparaparshinokh consisted in the alteration of what is called the 'subjective action' of three of its 'stopinders'. In one He lengthened the law-conformable duration, in another He shortened it, and in a third, disharmonized it.[9]

Splash—broken symmetry

Symmetries of reflection such as described above for the hexagon are called mirror symmetries, since a properly placed mirror would produce the transformation in question. Externally, the right and left halves of the human body are more or less mirror-symmetric. Louis Pasteur discovered that certain organic molecules, such as most of the amino acids that are the building blocks of proteins, come in two versions which are mirror symmetric: a laboratory chemical reaction which produces an amino acid produces equal amounts of "right-handed" and "left-handed" molecules. Yet in organic life on earth, only one form of the amino acids is present. Pasteur thought this fact held some kind of key about the nature of life, and its fundamental difference from inorganic materiality. While various explanations for this asymmetry have been put forth, including chance, none is clearly compelling, and Pasteur's mystery may still hold some secrets.

A further asymmetry of interest is the asymmetry between the two halves of the brain. The two cerebral hemispheres are more or less symmetric in mammals other than humankind. Animals do not show strong paw preferences, and lack symbolic language. In humans, language is primarily served by one hemisphere, while the other has specializations of its own. One could say that human beings, the most complex and the most conscious of creatures on earth, are also the most asymmetric, in a sequence of broken symmetries from the primordial unified force to the asymmetries of the two hemispheres, and speculate that this has some fundamental significance.

Symmetries and broken symmetries are intimately related with dimensionality. As noted above, in a totally symmetric void there would be no concept of space or time. While there are an infinite number of regular polygons in the plane, in three dimensions there are only five regular polygonal solids, the five platonic solids. As another example, rotational symmetry refers to a transformation within the dimensions of the entity in question; for instance the rotational symmetries of the hexagon are shown by rotating the hexagon within its plane. But mirror symmetries require a dimension above those of the entity: to show the mirror symmetry of the silhouette of a man, half has to be lifted above the plane and flipped over in order to be to superimposed upon the other half. The mirror symmetry of three-dimensional right and left hands requires a conceptual fourth dimension: the two hands cannot be superimposed in our space.

This peculiarity even applies to numbers. In the one-dimensional line, rational numbers would suffice, since any point could be specified as some fraction of the line. Irrational numbers arise because of the plane: the relation between the circle's diameter and its circumference, between the square's side and its diagonal. Cantor showed that although there are an infinite number of digits (1, 2, 3, 4, etc.) and an infinite number of rational numbers ($\frac{1}{2}$, $\frac{1}{3}$,

¼... ⅔... ¾, etc.), the rational numbers, though there seem to be more of them, can be put into one-to-one correspondence with the digits, and therefore they both belong to the same order of infinity. The irrational numbers, however, cannot be matched one-to-one with the rational numbers, so they comprise a greater infinity; the argument is simple and can be found in many books.[5] Similarly, a plane and a line are two different orders of infinity: there are an infinite number of points in a line, and an infinite number of lines in a plane. So these different infinities correspond to different dimensions. The rational numbers also exhibit a symmetry which the irrational numbers do not. The decimal expansion of a rational number is either exact (½ = 0.5), or a repeating decimal (⅓= 0.33333....). The most complex repeating decimals are those such as ⅐ (0.142857142857....) or 1/13 (0.076923 076923....) which nevertheless always repeat. Aside from the cyclic symmetry of repetition, these fractions also exhibit another kind of symmetry: the second half of the repeating series is complementary to the first half. For 0.142857, for example, 1 and 8 make 9, 4 and 5 make 9, and 2 and 7 make 9. This, when diagramed, produces mirror symmetric figures, such as the enneagram:

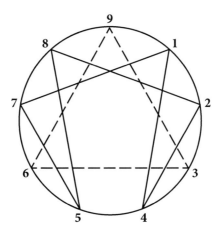

On the other hand, irrational numbers such as pi exhibit a different kind of symmetry, as seen earlier in the infinite series expansion of pi.

The concept of symmetry and broken symmetry has become very important in physics. Elementary particles have symmetry relationships, and such considerations have led to the discovery of new particles. Every charged particle, for instance, has a corresponding oppositely charged antiparticle. Yet in our universe, electrons and protons seem to predominate massively, with a paucity of positrons (positively charged electrons) or antiprotons: a broken symmetry.

Symmetry and broken symmetry in time are also fundamental. For time to be meaningful, there must be regularly occurring phenomena—cyclic, periodic events—by which time can be measured. The world is full of them, from atomic clocks to light, sound, and other kinds of waves, to the repetition of astronomical events: day and night, the cycles of the moon, the seasons of the year. The occurrence of periodic events breaks the symmetry of eventless time, but retains the symmetry of periodicity: moving forward or backward one cycle leaves a cyclic phenomenon unchanged. Much of classical physics was devoted to working out the mathematical laws underlying these periodicities.

These laws of physics have a time-reversal symmetry: if the revolutions of the earth were run backward in time, they would look like perfectly acceptable planetary motions, and this applies as well to the equations of wave motion. But in our everyday life things cannot run backward; the spilt milk never gathers itself up to return to the glass. This further broken symmetry is necessary for there to be possibilities and choices, along with the accompanying uncertainties and unpredictability. If there were no possibilities and no choices, there would be no need for intelligence.

In the nineteenth century, the study of irreversible phenomena led to the concept of entropy. The laws governing irreversible phenomena seemed to point to an inevitable accumulation of disorder, such that the universe would eventually become homogeneous and eventless. But if there were only possibilities without choices, there would be no need for intelligence either. Living things are by definition intelligent: up to a point, they make the choices that sustain themselves by maintaining their form, constantly taking in matter and energy and transforming them into what is needed. They maintain themselves far from equilibrium, or a state of uniform energy. When an organism dies, its intelligence no longer maintains its form, and its components revert towards equilibrium. This is largely an electrical phenomenon: every living cell maintains a potential difference across its membrane, and every chemical reaction involves a specific separation of charges.

In recent decades, inroads have been made into understanding these complex patterns that characterize living things. One major discovery in the mathematics of complex phenomena is that complex patterns can arise from simple causes. A number of factors seem to be required: time irreversibility, far-from-equilibrium structures, self-similarity on different scales, repeated feedback or iterations, and non-linearity.[10] Given these conditions, simple mathematical equations can produce a wide variety of complex patterns, which resemble those seen in living nature. Living things, conversely, conform to all of these conditions. Time is irreversible: all living things age and die. Life maintains itself far from equilibrium; the form is relatively stable while the material constituents are continually being replaced, with a steady supply of energy from the environment in the form of food, air, and sensory stimulation. Living things make extensive use of self-similar structures: the branchings of bronchial tubes, blood vessels and neurons allow interpenetration and intermixing of different substances needed for metabolism. Repeated and complex feedback is one of the most salient features of the living world, from the fact that everything eats and is eaten to the complex interactions that characterize social structures. And non-linearity, provided by the basic dimensions of the world, is the rule: a moment of passion can produce a new being, a step off the edge of a cliff can kill a person, and one incident can trigger a world war.

NOTES

1. P.D. Ouspensky, *In Search of the Miraculous: Fragments of an Unknown Teaching*, (New York: Harcourt, Brace, 1949), p. 127.

2. Richard Preston, "The mountains if pi," *The New Yorker*, March 2, 1992.

3. Quoted in Manfed Schroeder, *Fractals, Chaos, Power Laws: Minutes from an Infinite Paradise* (New York: W.H. Freeman and Co., 1991), p. 61.

4. John L. Casti, *Complexification: Explaining a Paradoxical World through the Science of Surprise* (New York: HarperCollins, 1994), p.139.

5. R. Penrose, *The Emperor's New Mind* (Oxford: Oxford University Press, 1989).

6. Quoted in James Gleick, *Chaos: Making a New Science* (New York: Viking, 1987), p. 273.

7. Ian Stewart, *Nature's Numbers: The Unreal Reality of Mathematical Imagination* (New York: Basic Books, 1995), p. 18.

8. Gregoire Nicholis and Ilya Prirogine, *Exploring Complexity: An Introduction* (New York: W.H. Freeman and Co., 1989), pp. 9–10.

9. G.I. Gurdjieff, *All and Everything. First Series. Beelzebub's Tales to his Grandson* (New York, Jeremy P. Tarcher/Penguin, 1992), p. 689–690.

10. A linear relationship is one in which the effect is directly proportional to the cause: "the number of apples equals 12 times the number of boxes" is a linear equation; "the volume of a cube equals its side cubed" is not. Thus, non-linearity arises from one dimension interacting with another: the apples and the boxes are of the same dimensionality, whereas the side and volume of a cube are not. Non-linearity and time irreversibility are intimately related, as in the examples at the end of the essay.

PYTHAGORAS IN 1999
A New Science of Mysticism

Pythagoras was the father of both Western science and Western mysticism. The historical Pythagoras is shrouded in legend—he was regarded as a god, and said to perform miracles—and there are many uncertainties about his life and teaching; even his dates (perhaps 569 B.C.E. to 470 B.C.E.) are uncertain. We know his thought through his many followers, including Plato and Plotinus, and both his life and thought through ancient biographers: Porphyry, Iamblichus, and Diogenes Laertius. It is fairly certain that he studied with contemporary Greek philosophers, and for prolonged periods in Egypt and Babylon. His teachings were largely kept secret during his lifetime, but there is little doubt that Pythagoras himself was an extraordinary individual. However, everything attributed to him is necessarily of uncertain origin, and much of what he taught may have come from more ancient sources.

What would Pythagoras have thought of our modern ideas about the world? Let us imagine that a time machine transported him to 1989, and he then studied the developments in mathematics, music, philosophy, and physical science of the last 2500 years. Now, ten years later, Parabola has had the opportunity to interview him.

CHRISTIAN WERTENBAKER: Well, sir, are you finding our modern civilization congenial?

PYTHAGORAS: Of course, your technological inventions are absolutely phenomenal and some are quite wonderful. And you have, at least in some places, developed the ideas of democracy and human rights that originated in my country over two thousand years ago. But your life is so hectic! What about the right to contemplate an idea for a long time, or to have an uninterrupted conversation? But we didn't come together to talk about that; you wanted to ask me about my scientific and mathematical ideas, and whether they make sense nowadays.

CW: You're right. I suppose the first question is: how has the nature of scientific and mathematical inquiry changed since you made your seminal contributions?

P: Let's take those two separately, although they are obviously related, and then we can also talk about what has *not* changed, which is what I find most interesting. First, the biggest change in science is pointed out in all your textbooks: it is the clarification of, and adherence to, what you call the scientific method—the idea that everything must be tested by experiment; speculation and logic are not enough. Of course, I did experiments with vibrating strings and other bodies, and we made careful observations of planetary motions, but back then experiment was not considered as important, nor did we have the sophisticated experimental tools that you have developed, the microscopes, telescopes, atom smashers, and so on. The precision of logic has also improved, mostly through the work of mathematicians. So, everything is more rigorous.

On the other hand, you could say—and many contemporary people believe this—that the soul, or feeling, has gone out of science. It's like your modern capitalism: making money is the priority, and quality, beauty, fulfillment, empathy are all secondary. Similarly, in science, producing facts has become paramount; the sense of meaning, of the nature and purpose of life, the sense of communion with a living, conscious universe, has gone out of the enterprise. We regarded numbers as divine principles, and connected with them not only through thought, but also through a higher feeling.

But this view—although there's some truth to it—is somewhat superficial. Many great scientists of your civilization were and are mystics, although they might not admit it. Newton spent more time on alchemy than he did on gravitation and the calculus. Einstein made statements like, "I want to know how God created this world. I am not interested in this or that phenomenon. I want to know His thoughts, the rest are details." And: "The creative principle resides in mathematics. In a certain sense, therefore, I hold it true that pure thought can grasp reality, as the ancients dreamed." [1] Many contemporary mathematicians and physicists feel, as Einstein did, that a mathematical theory can't be true unless it is also beautiful.

CW: Do you think that now we suffer from too much information? No one can take it all in, which makes it difficult to contemplate the bigger picture.

P: Yes, no one can even understand all of mathematics, and the mathematics is so complicated that it takes years to get a feel for it. In my day, an intelligent person could study all the knowledge there was. Now, one has to rely on experts to understand some scientific theories, and those experts may not grasp the meaning themselves. No one really understands the meaning of quantum theory, though some have tried very hard. For a long time—and still now to some extent—the view prevailed among physicists that it was pointless to try to understand the meaning; that's not what science is about. But humans can't help but look for meaning; it's their most compelling need. Much of modern physical and mathematical theory is contrary to the commonsense intuitions we develop living on the earth. Relativity and quantum theory, the cornerstones of modern physics, just don't make sense to us. But the truth of these theories, even if incomplete, can't be denied, and you all happily use your transistors, laser CDs, and atomic energy plants, none of which would exist if these theories weren't at least partially true and understood.

cw: It's interesting that now knowledge is disseminated freely—some would say too freely, given some of the horrors that science has produced—yet it remains hidden to many because of its complexity. In your day, knowledge was kept secret, even though it was more comprehensible.

p: Well, in a way it was more comprehensible. But again, this analysis is somewhat superficial. After all, one meaning of the word "mystical" is "hidden." The great mystical knowledge which has existed since very ancient times has always been, in part, contrary to ordinary common sense and inaccessible to the ordinary mind. Not that modern science and mysticism are the same thing. Their methods are very different. Science regards knowledge as external, in a sense: it has to be demonstrable by manipulations of the external world. Mysticism regards true knowledge as graspable from within, by a specially trained, more inclusive, higher consciousness. This presupposes that we humans can be in tune with the essence of the cosmos. For many scientists this is an unproved fantasy, and certainly people can claim all kinds of revelations which are demonstrably hallucinatory. So scientists demand external verification. But, in both cases, special training is needed, and when you get right down to it, faith in logic and observation also presupposes a kind of being in tune with the universe. In my more optimistic moments, I think that science came about in its present form in order to bring a different kind of rigor to mystical knowledge, and that the two kinds of knowing are destined to join together.

cw: Some think that quantum theory is closely related to mystical truths.

p: Yes. In quantum theory the kind of observation made determines whether an electron, or any of the other small building blocks of which things are composed, looks like a wave or a particle. One can interpret this as indicative of the inextricable role of consciousness in the universe. Many scientists would deny this interpretation, but some great ones, like von Neumann, came to just this conclusion. So maybe the end result of your modern science will be to confirm, from a different point of view, great, ancient mystical truths—though there's a long way to go, certainly—and get rid of some of the nonsense promoted in the name of mysticism. For that matter, I could do without some of the concoctions made out of my own ideas over the centuries.

I think the real difference between modern science and true mysticism is that the scientist deliberately tries to ignore the role of the subject in understanding the world. But no understanding exists except within a conscious being; the understanding does not exist on paper, in the formulas and diagrams. Science also does not consider differing capacities for understanding, dependent not just on intellectual training but on an even more rigorous development of a higher capacity for consciousness—the aim of the mystical teachings.

cw: What about mathematics? How has it changed? As you said, there is a lot more of it now, much of it completely obscure to most people.

p: The changes have been enormous, of course. It's like the difference between one of your Mercedes and an ancient Greek chariot. The basic principle of a moving conveyance on wheels, however, remains the same. In the case of mathematics, our saying, "everything is number," seems truer than ever. The amazing applicability of mathematical constructs to the physical world has been noted again and again. Now you have more kinds of numbers than we did. We didn't like to use even negative numbers, or zero, and were puzzled by what you now call irrational numbers, numbers like the square root of two, or pi, which we were aware of because of geometry, but didn't really consider to be numbers because they couldn't be expressed as exact ratios. Now all of these kinds of numbers, as well as complex numbers, involving the square root of minus 1, have been integrated into mathematics, with wonderful results. And it seems that there are no more kinds of numbers to find, so the process is complete.

One of the most successful discoveries is the calculus, which makes it possible to deal with smoothly changing processes, such as motion, by having a method of handling infinitesimally small changes. It gets around some of the paradoxes we came up with in ancient Greece, now known as Zeno's paradoxes. For instance, Zeno argued that motion was impossible because in order to get from point A to point B, it's first necessary to get to point C, in between the two, and before one can get to point C, one has to get to point D, between A and C, and so on, so that one can't actually ever get started. The calculus makes it possible to handle motion mathematically, including wave motion, which we understood only qualitatively.

cw: But hasn't mathematics then gotten away from what some regard as primitive notions of the applicability of simple whole numbers—one, two, three, etc.—to the workings of the universe?

p: Ah, no, you see it has merely added to these basic truths. There are still three dimensions of space, at least on our macroscopic scale, and this determines many things. It still takes a minimum of two points to determine a line, three to determine a plane figure like a triangle, and four to specify a solid, like a tetrahedron. These four numbers arranged in a triangle, and their sum, ten, were our holy tetraktys.

This tetraktys, by the way, is identical in its basic pattern to the ten components of Riemann's metric tensor, an essential part of the mathematics of Einstein's general theory of relativity.[2] I find this absolutely amazing.

The universe is still one whole. Your modern idea of everything beginning as a singularity which exploded in the Big Bang (you do have some amusing terms for things) means that everything is interconnected, although your science still does not have the view that the whole universe is alive. Many processes depend on two opposite forces interacting; it is this ever-perpetuating dynamic tension between opposites that creates the infinite multiplicity of phenomena. There are force and inertia, positive and negative charges, male and female. But it takes three forces or things interacting to produce an event or a phenomenon: all of your significant equations have three terms—force = mass x acceleration, energy = mass x the speed of light squared, electrical current = voltage/resistance—and if there are more terms in an equation, often the extra ones are constants. And look, there are three constituents of atoms: protons, neutrons, and electrons. There are three quarks in a proton or neutron, and quarks come in three "colors" and three pairs of "flavors." There are three notes in a basic chord, and three primary colors can mix to form any color. We have three internal modes of cognition: sensation, emotion, and intellect.

There are four dimensions, at least on our scale, three of space, and one of time. So events take place on a stage of four numbers. Five comes in with the golden section or golden mean, which interested us greatly. The formula for the golden ratio is $(\sqrt{5}+1)/2$. The golden section is found in the five-pointed star, and in other figures with pentagonal symmetry, and is related to the Fibonacci numbers. These numbers and the golden ratio show up prominently in the patterns of living things, from the arrangements of leaves on a stem, to the patterns of sunflower seeds and cacti, to the spirals of marine shells.[3]

DNA, seen on end, has ten-sided symmetry. Five and ten also show up in modern "theories of everything," as well as in Riemann's metric tensor—in so-called SU(5) symmetry, and some of these new "string theories," with ten dimensions. Our idea that ten was the only right basis for a number system— as opposed to twelve, or sixty, which are more convenient because they have more whole number divisors, and are used for some things, like clocks, and angles—may turn out to be profoundly true when modern physics gets a clearer picture of a real fundamental theory.

I could go on with the other numbers we found important, showing you both the continuing validity of our concepts and the connection between them and modern discoveries. The idea of "magic numbers," simple whole numbers that keep showing up everywhere, is very much alive.

In fact, modern quantum physics can be seen partly as a return to simple numbers. After a long development, following Newton, of the applications of continuous functions handled by the calculus, it became necessary to return to the discrete, when Planck and Einstein and others found that energy could only be emitted and absorbed in tiny discrete clumps or quanta. And this is where it gets most interesting. The basic paradox of quantum theory is that elementary things behave both as waves and as particles, as continuous things and as discrete things. This is similar to the behavior of vibrating strings, to which I devoted so much attention. Strings vibrate in a wavelike fashion, but only in discrete tones, the fundamental tone and overtones of the string, determined by the constraint of the ends of the string, which must remain relatively fixed. The string can only vibrate as a whole, in halves, in thirds, in fourths, etc., giving the fundamental tone, its octave, the fifth above that, the next octave, etc.

The description of atoms uses the same mathematical language: vibrating bodies with certain, now three-dimensional, overtones.

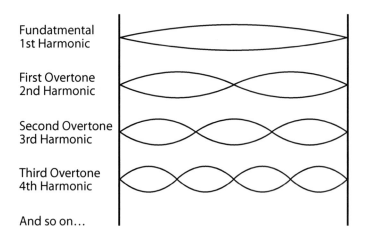

Fundatmental
1st Harmonic

First Overtone
2nd Harmonic

Second Overtone
3rd Harmonic

Third Overtone
4th Harmonic

And so on…

FIGURE 1 • Vibrational modes of a string

FIGURE 2 • Vibrational modes of hydrogen (electron probability distributions)

This creates your periodic table of the elements. Of course, now you have over a hundred elements instead of the earth, water, air, fire and ether which we called elements—these concepts though are still valid as representing not elements, but states of matter: solid, liquid, gas, plasma or electromagnetic energy, and what you now call the vacuum, which isn't as empty as that word implies, but teeming with energy. But isn't it striking that the modern elements fall into groups based on the number of electron orbitals, or vibratory modes, and that the orbitals can have, respectively, two times 1, 4, 9, and 16 electrons in them, the perfect squares, and these numbers in turn come from the simple whole numbers that determine the vibratory modes?

One thing is clear: everything vibrates, everything is in vibration, and the basic nature of the world is vibration. However abstract, complicated, and counterintuitive the modern theories become—whether it's ten or twenty-six dimensional strings vibrating, or abstract symmetry groups—this basic truth remains at the foundation of all of them. And the properties of vibrating things are determined by numbers, just as the musical tones are determined by numbers.

cw: But you felt that the musical scale, the simple notes represented by simple ratios:

1	⁹⁄₈	⁵⁄₄	⁴⁄₃	³⁄₂	⁵⁄₃	¹⁵⁄₈	2
do	re	mi	fa	sol	la	si	do

were the basic pattern on which the universe was built. Surely, things are not so simple, according to modern ideas.

P: No, they're not. But I think they're getting simpler as your theories develop. Actually, that was not the scale we used. In our scale some notes had different, though still relatively simple, ratios. And it's precisely this difficulty in determining the right scale to use which points out some most interesting aspects of the musical analogy.

Modern mathematical theories of the basic aspects of nature are based on the idea of symmetry. A symmetry exists when a property of a configuration of elements remains the same despite a manipulation of that configuration. For instance, if I rotate a hexagon a sixth of a turn, it still looks the same. If there were no symmetry, there would be no order, no predictability, no patterns; nothing recognizable could happen or exist. So there could be no intelligence, either. The most important advance in modern mathematical physics has been in the generalization of the idea of symmetry. The fact that I don't turn into an alligator when I walk down the hall reflects a symmetry: the laws of the universe don't change from one location to another. A similar symmetry exists for time: I don't vanish from one moment to another. But these symmetries are not perfect, because everything is not quite the same when I walk down the hall, nor from one moment to another. If there were perfect symmetry, nothing much could happen or exist, either. This is now called "broken symmetry." Symmetry and broken symmetry seem to underlie the basic organization of everything.[4]

CW: This seems to be more related to geometry than to numbers, or musical scales.

P: We regarded all of these areas as intimately related, and one of the beautiful aspects of modern mathematics is that it also has brought together many seemingly separate concepts. In fact, this is the antidote to the proliferation of information that we were talking about earlier.

Numbers have a kind of symmetry. Some number systems form what are now called mathematical *groups*, which I won't describe in detail. But the idea is that if you take, for instance, positive integers (1, 2, 3 etc.), and perform an operation on them by adding them together (just like you perform an operation on a hexagon by rotating it), the addition produces another integer, not a rabbit. On the other hand, if you divide two integers, you may get another integer (⁶⁄₂ = 3), or not (⅔ is not an integer). But ⅔ is a rational number—it's expressible as a ratio of whole numbers—so if you have a different group, the rational numbers, then you can divide them and always get a rational number. There are other requirements for groups, but that's one of the main ideas. So things can be categorized into various groups, which have various symmetries.

The notes of the musical scale, being produced by vibrations which are represented by simple ratios, have this kind of symmetry. If you go up a fifth (five notes in the scale, and $\frac{3}{2}$ the number of vibrations) from middle C, you get G, and if you go up a fourth from G ($\frac{4}{3}$ the number of vibrations), you get another C, an octave above the first one, because $\frac{3}{2} \times \frac{4}{3} = \frac{12}{6} = 2$, and the octave is twice the number of vibrations. In general, if you go up a certain number of notes in the scale, you get another note in the scale. But after a while there's a problem, which we saw long ago. For instance, if you go up twelve fifths:

C	G	D	A	E	B	
F#	C#	G#	D#	A#	E# or F	C

you get back to C, seven octaves above the first C. But this note is not quite the same as seven octaves, because $(\frac{3}{2})^{12}$—$\frac{3}{2}$ multiplied by itself 12 times—equals $^{531441}/_{4096} = 129.746$, whereas seven octaves is $(2)^7 = 128$. This difference $^{129.746}/_{128}$, or $^{531441}/_{524288}$ [*editor's note: which was named the Pythagorean comma*], is still a rational number, but certainly not a ratio of small numbers, like the notes of the scale. This problem led, a few hundred years ago, to your compromise scale, the tempered musical scale. And the problem exists because powers of 2 and powers of 3 never coincide exactly. So there you have a kind of imperfect symmetry, resulting simply from the properties of 2 and 3.

This slight difference between the twelfth fifth and the seventh octave, if they are played together, produces a slight dissonance, a tension, and creates another vibration. This can be regarded as a force between the two notes. I think this is analogous to the residual forces, resulting from imperfect symmetry, that create the different levels of interaction in the world. Let me quote from an article I read:

> The electromagnetic force binds electrons and nuclei to make atoms. The atoms, although they are electrically neutral, interact through a residual electromagnetic force to form molecules. The strong force binds quarks to make protons, neutrons, and all other hadrons, and the residual strong force between protons and neutrons is the so-called nuclear force that binds them into nuclei.[5]

Now, this is an analogy, not a direct correspondence. I don't claim that we anticipated the amazing advances that your civilization has made in discovering the mathematical underpinnings of the way the universe works.

But the astonishing applicability of mathematics to the physical world is even more obvious now than it was then. And simple numbers and ratios still play a major part. It may be, as further developments occur, that the correspondences between ancient and modern thought in this regard will become even more striking. Even now, I think your science is beginning to approach the realm of the ideal, the essence, that underlies all things, though it cannot be apprehended by science alone.

CW: What would you say is the major impediment to a more complete understanding?

P: The role of consciousness in the universe has to become part of your theories. Modern physics already has been forced to include the fact that the way in which a phenomenon is observed is an essential, though still mysterious, determinant of how reality manifests itself. But consciousness has to be explicitly put into the theory, and the relationship between the inner world of conscious beings and the outer world has to be understood. A great effort to understand consciousness is just starting now among brain scientists; maybe this will help lead to insights as to how these two worlds fit together.

NOTES

1. Michio Kaku, *Hyperspace: A Scientific Odyssey through Parallel Universes, Time Warps, and the 10th Dimension* (New York: Oxford University Press, 1994).

2. *Ibid.*, p. 41. The description of a curved space of four dimensions requires sixteen numbers:

$$
\begin{array}{cccc}
G_{11} & G_{12} & G_{13} & G_{14} \\
G_{21} & G_{22} & G_{23} & G_{24} \\
G_{31} & G_{32} & G_{33} & G_{34} \\
G_{41} & G_{42} & G_{43} & G_{44}
\end{array}
$$

six of which are redundant (because G12=G21, G13–G31, etc.), leaving ten. See also Julian Schwinger, *Einstein's Legacy: The Unity of Space and Time* (New York: Scientific American Books, Inc., 1986).

3. There are many books on these matters. See particularly H.E. Huntley, *The Divine Proportion: A Study in Mathematical Beauty* (New York: Dover Publications, Inc., 1970).

4. See "Nature's Patterns" in this volume.

5. Howard E. Haber and Gordon L. Kane, "Is Nature Supersymmetric?" *Scientific American*, June 1986, p. 52.

Some Thoughts on the Enneagram

Gurdjieff placed the enneagram symbol at the very center of his teaching, yet even for those who have spent years studying Gurdjieff's ideas it is one of its most incomprehensible aspects. He said:

"The enneagram is a universal symbol. All knowledge can be included in the enneagram and with the help of the enneagram it can be interpreted. And in this connection only what a man is able to put into the enneagram does he actually know, that is, understand. What he cannot put into the enneagram he does not understand. For the man who is able to make use of it, the enneagram makes books and libraries entirely unnecessary. *Everything* can be included and read in the enneagram. A man may be quite alone in the desert and he can trace the enneagram in the sand and in it read the eternal laws of the universe. And every time he can learn something new, something he did not know before.

"If two men who have been in different schools meet, they will draw the enneagram and with its help they will be able to establish at once which of them knows more and which, consequently, stands upon which step, that is to say, which is the elder, which is the teacher and which the pupil. The enneagram is the fundamental hieroglyph of a universal language which has as many different meanings as there are levels of men." [1]

Given this billing, those who have found in Gurdjieff's ideas a ring of truth difficult to find elsewhere must try to take the enneagram very seriously. Its resistance to comprehension indicates how far we really are from the level of understanding that Gurdjieff represented and embodied.

The enneagram is an integral part of many of Gurdjieff's movements, or sacred dances, and he said that "without taking part in these exercises, without occupying some kind of place in them, it was almost impossible to understand the enneagram.

"It is possible to experience the enneagram by movement. The rhythm itself of these movements would suggest the necessary ideas and maintain the necessary tension; without them it is not possible to feel what is most important." [2]

This clearly suggests that the intellect alone is not sufficient for understanding this symbol, or, as he elaborates elsewhere, for truly understanding anything at all. Nevertheless, the intellect is a component of any understanding, and its insufficiency does not justify abandoning intellectual rigor. "If you have not by nature a critical mind your staying here is useless." [3]

The published body of Gurdjieff's explanations of his ideas contains very little direct material on the enneagram, most of it in *In Search of the Miraculous*. Much of this material, though somewhat obscure, will be quoted here, since it forms an anchor for one's speculations.

This is how he introduced the enneagram to Ouspensky and the other Russian pupils:

"We have spoken earlier of the law of octaves, of the fact that every process, no matter upon what scale it takes place, is completely determined in its gradual development by the law of the structure of the seven tone scale...

"For uniting into one whole all knowledge connected with the law of the structure of the octave there is a certain symbol which takes the form of a circle divided into nine parts with lines connecting the nine points on the circumference in a certain order...

"This symbol takes the following form:

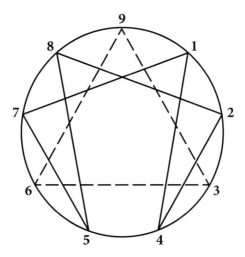

"The circle is divided into nine equal parts. Six points are connected by a figure which is symmetrical in relation to a diameter passing through the uppermost point of the divisions of the circumference. Further, the uppermost point of the divisions is the apex of an equilateral triangle linking together the points of the divisions which do not enter into the construction of the original complicated figure...

"The symbol which takes the form of a circle divided into nine parts with lines connecting them together expresses the law of seven in its union with the law of three.

"The octave possesses seven tones and the eighth is a repetition of the first. Together with the two 'additional shocks' which fill the 'intervals' mi-fa and si-do, there are nine elements.

"The complete construction of this symbol which connects it with a complete expression of the law of octaves is more complicated than the construction shown. But even this construction shows the inner laws of *one octave* and it points out a method of cognizing the essential nature of a thing examined in itself.

"The isolated existence of a thing or phenomenon under examination is the closed circle of an eternally returning and uninterruptedly flowing process. The circle symbolizes this process. The separate points in the division of the circumference symbolize the steps of this process. The symbol as a whole is *do*, that is, something with an orderly and complete existence. It is a circle—a completed cycle. It is the *zero* of our decimal system; in its inscription it represents a closed cycle. It contains within itself everything necessary for its own existence. It is isolated from its surroundings. The succession of stages in the process must be connected with the succession of the remaining numbers from 1 to 9. The presence of the ninth step filling up the 'interval' si-do, completes the cycle, that is, it closes the circle, which begins anew at this point. The apex of the triangle closes the duality of its base, making possible the manifold forms of its manifestation in the most diverse triangles, in the same way as the point of the apex of the triangle multiplies itself infinitely in the line of its base. Therefore every beginning and completion of the cycle is situated in the apex of the triangle, in the point where the beginning and the end merge, where the circle is closed, and which sounds in the endlessly flowing cycle as the two *do's* in the octave. But it is the ninth step that closes and again begins a cycle. Therefore in the upper point of the triangle corresponding to do stands the number 9, and among the remaining points are disposed the numbers 1 to 8.

"Passing on to the examination of the complicated figure inside the circle we should understand the laws of its construction. The laws of unity are reflected in all phenomena. The decimal system is constructed on the basis of the same laws. Taking a unit as one note containing within itself a whole octave we must divide this unit into seven unequal parts in order to arrive at the seven notes of this octave. But in the graphic representation the inequality of the parts is not taken into account and for the construction of the diagram there is taken first a seventh part, then two-sevenths, then three-sevenths, four-sevenths, five-sevenths, six-sevenths, and seven-sevenths. Calculating these parts in decimals we get:

$$\frac{1}{7} \quad 0.142857\ldots$$
$$\frac{2}{7} \quad 0.285714\ldots$$
$$\frac{3}{7} \quad 0.428571\ldots$$
$$\frac{4}{7} \quad 0.571428\ldots$$
$$\frac{5}{7} \quad 0.714285\ldots$$
$$\frac{6}{7} \quad 0.857142\ldots$$
$$\frac{7}{7} \quad 0.999999\ldots$$

"In examining the series of periodic decimals obtained we at once see that in all except the last the periods consist of exactly the same six digits which run in a definite sequence, so that, knowing the first digit of the period, it is possible to reconstruct the whole period in full.

"If we now place on the circumference all the nine numbers from 1 to 9 and connect those numbers which are included in the period by straight lines in the same sequence in which the numbers stand in the period, according to which number we start from, we shall obtain the figure found inside the circle. The numbers 3, 6, and 9 are not included in the period. They form the separate triangle—the free trinity of the symbol.

"Making use of 'theosophical addition' and taking the sum of the numbers of the period, we obtain *nine*, that is, a whole octave. Again in each separate note there will be included a whole octave subject to the same laws as the first. The positions of the notes will correspond to the numbers of the period and the drawing of an octave will look like the following:

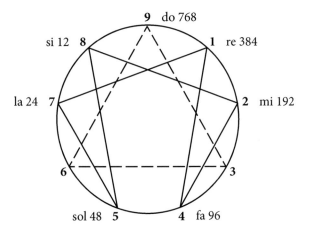

"The triangle 9-3-6, which unites into one whole the three points on the circumference not included in the period, connects together the law of seven and the law of three. The numbers 3-6-9 are not included in the period; two of them, 3 and 6, correspond to the two 'intervals' in the octave, the third is, so to speak, superfluous and at the same time it represents the fundamental note which does not enter the period. Moreover, any phenomenon which is able to act reciprocally with a phenomenon similar to it sounds as the note do in a corresponding octave. Therefore do can emerge from its circle and enter into orderly correlation with another circle, that is, play that role in another cycle which, in the cycle under consideration, is played by the 'shocks' filling the 'intervals' in the octave. Therefore, here also, by having this possibility do is connected by the triangle 3-6-9 with those places in the octave where the shocks from outside sources occur, where the octave can be penetrated to make connection with what exists outside it. The law of three stands out, so to speak, from the law of seven, the triangle penetrates through the period and these two figures in combination give the inner structure of the octave and its notes."[4]

At this point two comments are in order. First, the "law of seven" or the "law of octaves," and the "law of three" are discussed in great detail in *In Search of the Miraculous* and elsewhere, and cannot be expounded on here. Second, the "theosophical addition" referred to is the method of adding the digits of any multi-digit number, and then adding the digits of the resulting number, and so on, until a single digit is obtained. For example, 21 becomes 3, 49 becomes 13 and then 4, 286 becomes 16 and then 7. Equivalently, one can "cast out nines," that is, subtract nine repeatedly until a single digit is left. This method is useful, among other things, in checking the results of any arithmetic calculation. For example:

$$283 \times 486 = 137538$$

To check the result, the two multiplicands are reduced to their single digits by "theosophical addition," i.e. to 4 and 9, the two digits are multiplied to make 36, the single digit of which is 9, and the result of the original multiplication should also reduce to 9, otherwise there is an error. In the days before calculators, this was a useful tool.

In trying to understand the enneagram, one starting place is to inquire into its numerology. For, if the enneagram represents universal laws, either the number pattern itself is significant and a clue to these laws, or the number pattern is a fortuitous accident, which just happens to produce a figure which can serve as a diagram or symbol of these laws. Gurdjieff suggests the former is the case: "The laws of unity are reflected in all phenomena. The decimal system is constructed on the basis of the same laws." This statement, and the formation of the enneagram in the decimal system, suggest that the decimal system is not an arbitrary choice, possibly based on the number of our fingers, but that there is something more fundamental about it.

It is first worthwhile to see whether the decimal expansion of the fraction $\frac{1}{7}$ is unusual within the decimal system.

$$\frac{1}{2} = 0.5$$
$$\frac{1}{3} = 0.3333....$$
$$\frac{1}{4} = 0.25$$
$$\frac{1}{5} = 0.2$$
$$\frac{1}{6} = 0.16666....$$
$$\frac{1}{7} = 0.142857142857142857....$$
$$\frac{1}{8} = 0.125$$
$$\frac{1}{9} = 0.1111....$$

For the first 9 digits, 2, 4, 5, and 8 produce exact results; 3, 6, and 9 produce single digit repeating decimals; and the pattern produced by 7 is quite unique. Going beyond 9:

$\frac{1}{10}$ = 0.1

$\frac{1}{11}$ = 0.090909. . . .

$\frac{1}{12}$ = 0.083333. . . .

$\frac{1}{13}$ = 0.076923 076923. . . .
    another complex repeating decimal, of six digits.

$\frac{1}{14}$ = 0.07142857 142857. . . .
    the same repeating pattern as $\frac{1}{7}$, as one might expect.

$\frac{1}{15}$ = 0.06666. . . .

$\frac{1}{16}$ = 0.0625

$\frac{1}{17}$ = 0.0588235294117647 05882352941176470. . . .
    another complex repeating pattern, with sixteen digits.

$\frac{1}{18}$ = 0.05555. . . .

$\frac{1}{19}$ = 0.052631578947368421 052631. . . .
    an eighteen-digit repeating pattern.

$\frac{1}{20}$ = 0.05

$\frac{1}{21}$ = 0.047619 047619. . . .

$\frac{1}{22}$ = 0.0454545. . . .

$\frac{1}{23}$ = 0.0434782608695652173913 04347826. . . .
    a 22-digit repeating pattern.

$\frac{1}{24}$ = 0.0416666. . . .

$\frac{1}{25}$ = 0.04

$\frac{1}{26}$ = 0.0384615 384615. . . .

$\frac{1}{27}$ = 0.0370370. . . .

$\frac{1}{28}$ = 0.03571428 571428

$\frac{1}{29}$ = 0.0344827586206896551724137931 03448. . . .
    a 28-digit repeating pattern.

There are many interesting features of these numbers, long familiar to mathematicians. All fractions in decimal form are either exact, or produce repeating patterns, unlike numbers like $\sqrt{2}$, or π, or e, or phi (about which more later), which produce an endless series of digits without apparent pattern. These numbers, however, are expressible as infinite series of fractions, with beautiful patterns:

$$e = 1 + 1 + \tfrac{1}{2}! + \tfrac{1}{3}! + \tfrac{1}{4}! \ldots$$
$$\tfrac{\pi}{4} = 1 - \tfrac{1}{3} + \tfrac{1}{5} - \tfrac{1}{7} + \ldots$$

It is logically necessary that the decimal expansion of a fraction produce either an exact number, or a repeating pattern. If a number is divided by another, either there is no remainder, in which case the result is exact, or there is a remainder. When the remainder times 10 is again divided by the denominator, to arrive at the next decimal place, the same two possibilities exist. Eventually, one arrives at no remainder, in which case the result is exact; or at a remainder that is the same as one of the previous remainders, in which case the digits repeat themselves endlessly.

Other interesting features of these numbers are: 1) the complex repeating patterns often have a number of digits in the pattern that is one less than the original divisor. 2) Often the sequence can be divided in two: each digit in order from the start of the sequence is the complement (adding up to 9) of each digit in order beginning from the middle.

In any case, the pattern formed by the fraction $\tfrac{1}{7}$ is unique among the first 9 digits, and though it is only the first of a series of complex repeating patterns produced by fractions, it is the most striking, concise, and beautiful, in its inclusion of all the digits except for 3, 6, 9, and 0, each appearing only once.

If this sequence has some universal meaning, it might appear elsewhere. But to begin with, is there anything about the number 10 that singles it out as a non-arbitrary base for a number system? Aside from our fingers and toes, the numbers 10 and 5 appear in nature in the pentagonal symmetry of certain flowers, and in starfish. The DNA molecule, seen on end, forms a ten-sided figure. Two of the five platonic solids, the icosahedron and the dodecahedron, have pentagonal patterns.

One of the most intriguing places where the number 5 turns up repeatedly is in connection with the golden mean, or phi. The golden mean, or golden section, refers to the ratio that results when a line is divided into two segments, such that the ratio of the length of the whole line to the length of the longer segment is the same as the ratio of the length of the longer segment to the length of the shorter segment.

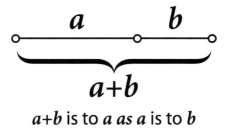

a+b is to *a as a* is to *b*

The formula for this ratio, for a shorter segment set equal to 1, i.e. $b = 1$, is: $(a+1)/a = a/1$, or $a^2 = a + 1$, or $a^2 - a - 1 = 0$, and $a = (1+\sqrt{5})/2$, called phi, or $(1-\sqrt{5})/2$, called phi'.

The golden section has been a source of fascination since ancient times, appearing in geometry (notably in the pentagonal star, symbol of the Pythagoreans), in art, and in a variety of places in nature, as detailed in a number of books (e.g. 5 and 9). By repeatedly applying the ratio, self-similar growing patterns are formed:

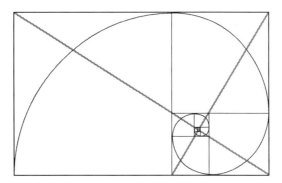

One of the most beautiful of these is the logarithmic spiral, found in certain shells, notably the Nautilus.

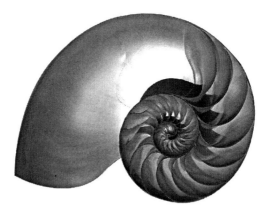

Thus, the golden section ratio is a means for growing organisms to retain the same shape as they grow, which is perhaps one reason it is so ubiquitous.

Intimately related to the golden mean is the Fibonacci series of numbers, formed by adding digits in sequence to produce the next:

$$1 + 1 = 2,\ 2 + 1 = 3,\ 3 + 2 = 5,\ 5 + 3 = 8,\ \text{etc.}$$

giving the sequence:

1, 1, 2, 3, 5, 8, 13, 21, 34, 55, 89, 144, 233, etc.

The ratios of adjacent numbers converge on phi:

1:1=1, 2:1=2, 3:2=1.5, 5:3=1.666, 8:5=1.6, 13:8=1.625, 21:13=1.6153846, 34:21=1.6190476, 55:34=1.617647, 89:55=1.6181818, 144:89=1.6179775, 233:144=1.6180555. Phi=1.61803349899887...

The Fibonacci numbers are also found in nature, such as the opposing spiral patterns of sunflowers and some cactuses:

Counting the spirals in one direction and those in the other direction, one finds two adjacent Fibonacci numbers.

The Fibonacci numbers also appear in the branching patterns formed by following certain rules, which apply to the genealogy of drone bees, for instance, and to the pattern of branches of a Sneezewort.[5]

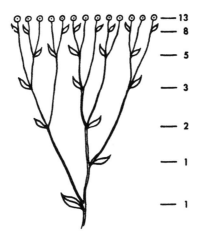

The rule is that any given branch can divide in two either after one generation or two, and one of the two branches of a division will divide again after one generation, and the other after two.

If we take the Fibonacci series, and reduce each multi-digit number by the method of theosophical addition, we get the following:

1	1	987	6
1	1	1597	4
2	2	2584	1
3	3	4181	5
5	5	6765	6
8	8	10946	2
13	4	17711	8
21	3	28657	1
34	7	46368	9
55	1	75025	1
89	8	121393	1
144	9	196418	2
233	8	317811	3
377	8	514229	5
610	7	832040	8 etc

(Note that $832040/514229 = 1.618033989$ is even closer to Phi than $233/144$.)

The reduced sequence thus formed can be arranged in two rows of twelve numbers, after which it repeats itself:

112358437189
887641562819
112358437189
887641562819

This series has several interesting properties. First is the fact that after twenty-four numbers it repeats itself. Second is the arrangement of the digits: the numbers 3, 6, and 9 occur in a regular pattern, every fourth digit. The other digits—1, 2, 4, 5, 7, 8—are interspersed in between. This is like the pattern of the enneagram, except that the non-multiples of three are not in the same order as in the enneagram, and they come in groups of three rather than pairs. Third, the two rows of twelve numbers form complements: each vertical pair adds up to 9, except for the last two 9s, which, by adding up to 18, reduce to 9 as well.

It is interesting that Leonardo of Pisa, alias Fibonacci, besides introducing Fibonacci numbers to Europe eight hundred years ago, also brought Arabic numerals and the decimal system.

Gurdjieff defines the law of three as follows:

"A new arising from the previously arisen through the 'Harnel-Miaznel,' the process of which is actualized thus: the higher blends with the lower to actualize the middle and thus becomes either higher for the preceding lower, or lower for the succeeding higher. . ."[6]

A division can be regarded as an example of the law of three: one number (higher) is divided by another (lower), which results in a third (middle, which needn't be between the first two numbers, although a re-arrangement of the three can put them in that order). In a division with a remainder, the remainder multiplied by 10 can be regarded as a transient "middle," which is divided again by the original divisor, etc., until a result emerges, either exact or a repeating series. A division which has a series of remainders can then be seen as an internal sequence of triads which eventually produces the overall triad of dividend, divisor, and quotient.

In this connection it is worth remembering that each digit of the quotient 0.142857 (or any other quotient) is on a scale that is ten times smaller than that of the previous digit. So the enneagram could be pictured like this:

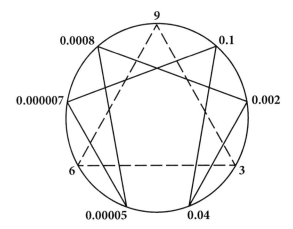

This suggests that the interactions symbolized by the enneagram are interactions on different scales, or levels. Gurdjieff says:

"We have spoken earlier of the law of octaves, of the fact that every process, no matter upon what scale it takes place, is completely determined in its gradual development by the law of the structure of the seven-tone scale. In connection with this it has been pointed out that every note, every tone, if taken on another scale is again a whole octave. The 'intervals' between mi and fa and between si and do which cannot be filled by the intensity of the energy of the process in operation, and which require an outside 'shock,' outside help so to speak, connect by this very fact one process with other processes. From this it follows that the law of octaves connects all processes of the universe and, to one who knows the scales of the passage and the laws of the structure of the octave, it presents the possibility of an exact cognition of everything and every phenomenon in its essential nature and of all its interrelations with phenomena and things connected with it."[7]

In an extremely lucid monograph on the nature of physical law, David Bohm[8] returns repeatedly to the necessity of looking at the interrelations between various levels of phenomena—e.g. the weather is made of gases, made of molecules, made of atoms, subatomic particles, etc.—but not simply as one level of phenomena being composed of things on another level, and determined by the properties of those things, until we find a fundamental smallest level which determines everything; but rather as a set of reciprocal relationships, whereby the lower level affects the higher and vice versa, so that reductionism is not possible. "Up to the present, the various kinds of things existing in nature have, at least as far as investigations in the field of physics are concerned, been found to be organized into levels. Each level enters into the substructure of the higher levels, while, vice versa, its characteristics depend on general conditions in a background determined in part in other levels both higher and lower, and in part on the same level."[8]

In his discussions of the law of seven, or the law of octaves, Gurdjieff discussed one phenomenon in detail in relation to this law: the metabolism of food, air, and sensory impressions (which he regarded as three levels of food) by the organism. These three levels correspond to his division of the organism into three parts, governed by three "brains": the physical, emotional, and intellectual aspects of man. The "food octave" produces an enneagram, like this:

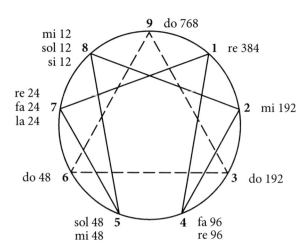

At this point I will certainly have lost the reader who is not already quite familiar with Gurdjieff's ideas. To clarify a bit: the numbers 768, 384, etc., refer to levels of materiality. Smaller numbers refer to finer materiality: *do* 768 is ordinary food, *do* 192 is air, *do* 48 is the food of sensory impressions. The metabolism of these foods is the process of coarser substances being transformed into finer ones, e.g. ordinary food is transformed into energy for the body's functioning. This process of metabolism is represented by an ascending scale or octave. The three kinds of food interact, so that the octave of ordinary food requires a "shock" to continue to evolve when it reaches the interval *mi-fa*; this corresponds to the entry of air, which is required (as oxygen) for the full conversion of the products of metabolism into energy that can be used by the body. See also "The Cosmic Metabolism of Form."

The enneagram, among its other remarkable mathematical properties, is bilaterally symmetrical. Correspondingly, the right half of the enneagram for the metabolism of food corresponds in a quite direct way to the scientifically known steps of metabolism. The left half, however, deals with substances that we do not ordinarily regard as physical, beginning with substance 48, corresponding to "impressions." According to Gurdjieff, we do not digest impressions at all, ordinarily. All of the complex coordinated acts we perform, guided by our sense impressions, all the turning thoughts and memories which make up our ordinary mental life, are regarded as serving a mechanical level of brain activity. The real metabolism of impressions requires, at the moment of entry of the impression, another shock—a conscious one, in which the perceiver is simultaneously aware of the impression and of himself receiving it. This shock allows the impressions octave to begin, and the air octave to pass from *mi* to *fa*.

The left half of the enneagram thus corresponds to our inner life, the right half to our physical body. The enneagram shows the dichotomy between the two: their opposition, symmetry, and interaction. The emotions can be seen as straddling the two. The interaction of bilateral symmetry, or dichotomy, and the triangle, shows us a difficult-to-comprehend relationship between two and three.

The bilateral symmetry of the complex figure within the enneagram corresponds to the symmetry of the number 142857, the first three digits and the last three forming pairs that each add up to 9. This results from the pattern of the remainders: when the remainder 6 is reached, one less than the 7 that would, divided by 7, produce unity, the complementary sequence of digits and remainders is produced, until the remainder returns to 1 and the whole sequence starts again.

$$
\begin{array}{r}
0.1428571 \\
7\overline{)10} \\
30 \\
20 \\
60 \\
40 \\
50 \\
10
\end{array}
$$

In the enneagram, each half of the sequence forms a triad. On the right, *re*, interacting with *fa*, and with the help of the shock, produces *mi*. This corresponds to the interactions of the three-carbon metabolites of food with appropriate enzymes and oxygen, to produce ATP and other high-energy compounds which then provide the energy for all the activities of the body, including those of the brain. On the left, the sequence 857 is symmetrical with the sequence 142 on the right, but since the outer sequence of notes is moving up instead of down on the circle, there is a reversal: the origin of the triad is its finest substance rather than its coarsest. This corresponds to the difference in the nature of the shock: mechanical and automatic on the right, conscious and intentional on the left.

It is intriguing then that the reduced Fibonacci series, which has a separation between the digits 124578 and the digits 369 reminiscent of the enneagram, also has a pattern of complementary sequences. It is worth noting that the end of the first half of the sequence, the twelfth Fibonacci number, is 144, or the square of 12. However, the 24th number, 46368, does not have an exact square root, cube root, or any root up to tenth, and is not, to my perception, an interesting number. Nor can I see how the complimentary second twelve Fibonacci numbers correspond in any way to a return, or a symmetry, in the places in nature where Fibonacci numbers appear. For that matter, Fibonacci numbers in nature don't seem to go beyond the first thirteen (a sunflower can have 144 and 233 opposite spirals).[9]

Similar patterns can be generated with other series:

Multiples of 2: 2, 4, 6, 8, 10, 12, 14, 16, 18, 20, 22, 24, 26, 28, 30

Theosophical sums: 2, 4, 6, 8, 1, 3, 5, 7, 9, 2, 4, 6, 8, 1, 3

These form a repeating pattern after nine numbers, with a similar segregation of 3, 6, 9 from the other digits.

Multiples of 4: 4, 8, 12, 16, 20, 24, 28, 32, 36, 40, 44, 48, 52

Theosophical sums: 4, 8, 3, 7, 2, 6, 1, 5, 9, 4, 8, 3, 7

These do likewise.
Multiples of 3 or 6, however, only give 3s, 6s, and 9s.

3, 6, 9, 12, 15, 18, 21, 24, 27, 30

3, 6, 9, 3, 6, 9, 3, 6, 9, 3

One can see (and mathematicians can prove) that any series of integers that is generated by a fixed rule, involving arithmetic operations on the integers, will produce theosophical sums that eventually repeat themselves. This is true in any number system (to any base, including 10). For instance, the Fibonacci series in base 2 repeats itself after three digits, in base 3 after eight, in base 4 after six, in base 5 after 20, in base 6 after 24, in base 7 after 16, in base 8 after 12, etc.[10]

The geometric series produced by repeated doubling, which also corresponds to the serial octaves of a note, gives:

1, 2, 4, 8, 16, 32, 64, 128, 256, 512, 1024, 2048, 4096

1, 2, 4, 8, 7, 5, 1, 2, 4, 8, 7, 5, 1

This is a series containing *only* the digits other than 3, 6, 9, and 0, like the repeating decimal expansion of 1/7. This is of particular interest because this series corresponds to both serial octaves, e.g. (multiplying the numbers by 100 to be more musically realistic) 100 cycles per second (cps), 200 cps, 400 cps, etc., and the numbers of overtones of a note between one octave and the next. The note 100 cps has overtones consisting of 200 cps, 300 cps, 400 cps, etc., because a string of fixed length and tension can vibrate as a whole, or in halves, or in thirds, etc.

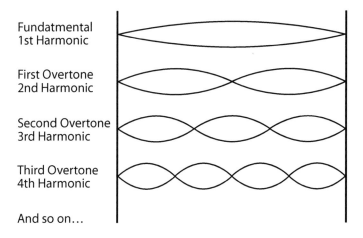

Fundatmental
1st Harmonic

First Overtone
2nd Harmonic

Second Overtone
3rd Harmonic

Third Overtone
4th Harmonic

And so on…

Similar rules apply to vibrations in a tube, as in wind instruments. Any real note produced by an instrument is always a mixture of the fundamental tone (the string vibrating as a whole) and the overtones, which gives notes their timbre, or quality. Different combinations of strength of the various overtones, determined by the characteristics of the particular instrument, give different timbres to the notes, making instruments recognizable by the quality of their tones.

The overtone series for the note 100 cps (call it *do*)

100	*do*	1700	
200	*do*	1800	*re*
300	*sol*	1900	
400	*do*	2000	*mi*
500	*mi*	2100	
600	*sol*	2200	
700	between *la* and *si*	2300	
800	*do*	2400	*sol*
900	*re*	2500	
1000	*mi*	2600	
1100		2700	
1200	*sol*	2800	
1300		2900	
1400		3000	*si*
1500	*si*	3100	
1600	*do*	3200	*do*

consists of octaves, the *dos*, with increasing numbers of notes in between, so that between one octave and the next there are (including the higher octave) 1, 2, 4, 8, 16, etc. notes. Many of these notes (as shown above) are notes in our scale, since the notes of the scale are derived from ratios: $\frac{9}{8}$ of the vibration rate of *do* produces *re*, $\frac{5}{4}$ (or $\frac{10}{8}$) *mi*, $\frac{4}{3}$ *fa*, $\frac{3}{2}$ (or $\frac{12}{8}$) *sol*, $\frac{5}{3}$ *la*, $\frac{15}{8}$ *si*, $\frac{16}{8}$ (or double) *do*.

There is no *fa* or *la* in the overtone series, since there is no overtone in a $\frac{4}{3}$ or $\frac{5}{3}$ relationship to any of the *dos*. Ratios involving 3 in the denominator stand out as different from ratios involving multiples of 2, just as in the various repeating patterns of numbers multiples of 3 stand out from the other digits.

In the series formed by doubling, as well as that produced from ⅟7, multiples of 3 are not present, just as *fa* and *la* are not present in the harmonic series. In the enneagram related to the musical scale, the notes *fa* and *la* follow the "gaps" in the series where the triangle (points 3 and 6) inserts itself. These gaps, or "intervals" as Gurdjieff calls them, are of fundamental importance in his scheme, being places where a shock is necessary for a process to continue.

However, there are notes in the overtone series that are in a ⅘ or ⅗ relationship to the note *sol: do* and *mi*.

Another pair of series illustrates the fundamental separation between two and three: powers of two: 2, 4, 8, 16. . .—repeated doublings—contain only the "theosophically reduced" digits 1, 2, 4, 5, 7, 8, while powers of three: 3, 9, 27, 81, 243. . . have only the reduced digits 3, 6 and 9. Consequently powers of two and powers of three never meet. This leads to the musical conundrum that resulted in the compromise tempered musical scale, and may have other implications with regard to the relationship between levels (see "Pythagoras in 1999").

In the view of Pythagoras, Plato, and many others, the properties of the universe derive in a direct and fundamental way from the properties of simple mathematical objects. It is often remarked that there is an uncanny correspondence between the physical world and mathematical concepts, and seemingly arbitrary pure mathematical derivations often are found years later to perfectly describe some corner of physical law. With this in mind, one could consider a fundamental dichotomy in both the physical universe and mathematical objects: the dichotomy between the continuous and the discrete. One could say that if there were not continuity, there would be no connection between things, and if there were not discreteness, there would be no separate entities.

> *You and I are one.*
> *Yet you are you, and I am I.*
> *How can that be?*
> *Let it be!*[11]

This dichotomy appears in an irreducible way in quantum physics, in which the fundamental constituents of matter and energy—photons, electrons, quarks—must be viewed both as continuous waves, and as discrete particles. They show their wavelike aspects when propagating in space, and their particle-like aspects when interacting with matter.

The line without ends in one dimension, the plane without boundaries in two dimensions, and endless space in three dimensions, are simply the space in which things occur, or the continuum. In the plane, the fundamental smooth, continuous object is the circle. Steady movement of a point around a circle, seen end-on from a viewpoint perpendicular to the plane of the circle, produces a sine wave, which is the basic form of wave (see Figure 1 in "Does Man Have Three Brains?"). All waves can be regarded, by Fourrier analysis, as sums of simple sine waves, of different frequencies and/or phases. So the simple continuous mathematical object, the circle, determines the properties of waves, and everything is, in one aspect of reality, composed of waves.

When a line acquires ends, it becomes discrete. A string stretched between two points is a physical manifestation of this. Such a string can vibrate within its own plane. Vibrating, it produces sine waves. Being bounded at both ends, its modes of vibration are restricted to multiples of its fundamental frequency. Here, the properties of the integers come into play, as detailed above for the harmonic series. One of the properties of the integers is that multiples of 3 segregate out from the other numbers. One could regard the inner figure of the enneagram made by 142857 to represent discrete phenomena, and the triangle made by 369 to represent continuousness. Perhaps the obscure statement by Gurdjieff, "The apex of the triangle closes the duality of its base, making possible the manifold forms of its manifestation in the most diverse triangles, in the same way as the point of the apex of the triangle multiplies itself infinitely in the line of its base," is related to this.

As described in detail elsewhere (see "Does Man Have Three Brains?"), one can postulate three dimensions of time: the ordinary linear time we are familiar with, moving steadily from past to future at "one second per second"; the perpetual existence of each moment—eternity; and the realm of possibilities. The dimension of ordinary time is represented by a line, that of eternity by a wave, that of possibility by a tree. These dimensions can be found in the enneagram: ordinary time is the movement around the circle from 1 to 9, eternity is in the triangle 369, and possibility is in the figure 142857.

Notes

1. P.D. Ouspensky, *In Search of the Miraculous: Fragments of an Unknown Teaching* (London, Routledge & Kegan Paul LTD, 1950), p. 294.

2. Ouspensky, p. 295.

3. G.I. Gurdjieff, *Views From the Real World* (New York, E.P. Dutton & Co. Inc., 1973), p. 283.

4. Ouspensky, pp. 285–290.

5. H.E. Huntley, *The Divine Proportion: A Study in Mathematical Beauty* (New York, Dover Publications, Inc., 1970), pp. 160–163.

6. G.I. Gurdjieff, *All and Everything. First Series. Beelebub's Tales to his Grandson* (New York, E.P. Dutton & Co. Inc., 1964), p. 751.

7. Ouspensky, p. 285.

8. David Bohm, *Causality and Chance in Modern Physics* (London, Routledge & Kegan Paul Ltd., 1957), p. 140.

9. Gyorgy Doczi, *The Power of Limits* (Boulder and London, Shambala, 1981), p. 5.

10. Thanks to Rick Sharpe; there is also an article "On the periods of the Fibonacci sequence modulo M", by Amos Ehrlich, in the Feb. 1989 *Fibonacci Quarterly*, pp. 11–13. (Modulo M is the mathematical terminology for reducing numbers by casting out M's.)

11. Tilo Ulbricht, personal communication.

VIBRATIONS:
THE UNIVERSAL
MEDIUM OF EXCHANGE

VIBRATIONS

G.I. Gurdjieff made an important contribution to the understanding of the universe and humanity's place in it, presenting what he said was very ancient knowledge in a modern form "in agreement with the form of mentation now established among contemporary people."[1] An aspect of this was the inclusion of a scientific viewpoint in his expositions. He famously said that everything is material, even God, but that materiality had many different degrees of fineness. This is obvious even on an ordinary level: metal is denser than wood. When we go to even more dramatic differences in materiality, comparing, for instance, solids, liquids, and gases, we are not just speaking of differences in density, but of changes in behavior, even dimensionality, as well. For instance, a solid tends to stay together as an entity as it moves about, whereas liquids change shape as they flow along surfaces, and gases disperse in three dimensions. Beyond these levels, fire is also material, a plasma. Light and other electromagnetic vibrations, like heat, radio waves, x-rays and cosmic rays, are also properly considered as material. Physics has shown that phenomena once considered to be exclusively vibrations also require a description as particles: electromagnetic vibrations are one aspect of light, photons another. The reverse holds true as well: atoms and elementary particles, the building blocks of matter, can also be described as vibrations. So it is not so radical now to regard thoughts, emotions, the soul, and even God as material.

The most interesting and important part of Gurdjieff's teaching is related to vibrations, and it seems to me that since his time his views have been increasingly validated by science.

Gurdjieff said:

"It is necessary to regard the universe as consisting of vibrations. These vibrations proceed in all kinds, aspects, and densities of the matter which constitutes the universe; they issue from various sources and proceed in various directions, crossing one another, colliding, strengthening, weakening, arresting one another, and so on."[2]

It is now quite clear that everything vibrates, from electrons and photons, to atoms, molecules, cells, bodies, planets, and stars. These entities vibrate in many frequency ranges, and interact by means of vibrations, most prominently electromagnetic vibrations. To interact, things must be tuned to each other in some way. There are patterns, harmonic relationships, of all kinds. And ultimately, when science probes the smallest constituents of matter, nothing is left except equations describing vibrations.

Vibrations are essentially regularities in time. If there was no regularity in time, there could be no enduring entities. The basic structure of the world, with discrete objects and living beings in it, maintaining their integrity for a certain duration, depends on regularity in time. But the regularity is not fixed, or nothing could happen, or change. So there is a constantly changing, developing, interacting pattern of vibrations.

Another statement from Gurdjieff:
"The reception of external impressions depends on the rhythm of the external stimulators of impressions and on the rhythm of the senses."[3]

After this sentence, Gurdjieff goes on to describe how different components of sound are meant for different brains or "centers"—the main ones being the intellectual, the emotional, and the instinctive/moving. If the listener's organism is not tuned properly, parts of the sound meant for a particular center will not get there, and the reception of an impression will be incomplete or distorted. A simple example of this occurs in listening to another person speak. There are several aspects to the speech: the contents of what is being said, which is perceived intellectually, the emotional tone in which it is being said, and the sensations produced on the body by the sound. It is quite common for two people to disagree because they are reacting to each other's tone, even though on examination they would agree with each other's thoughts. And vice versa: they might agree with each other because their tones of voice are compatible on an emotional level, but a closer look might reveal that their thoughts are quite contradictory. The different components of the vibrations of speech are not being properly distinguished and processed.

There seems to be a paradigm shift developing in neuroscience, having to do with vibrations. For many years, scientists and physicians have recorded brain waves, from the scalp or the surface of the brain, which are electromagnetic vibrations generated by the activity of large groups of brain cells. The predominant frequencies vary according to the state of the organism: sleep, dreaming sleep, waking, relaxing, concentrating, meditation. The new realization is that these vibrations also relate to the *contents* of consciousness, not just to its state. There are many parts of the nervous system, which perform many functions: brain areas devoted to the analysis of visual stimuli—with subareas for the processing of color, shape, visual movement, depth and so on— areas specialized in analyzing sounds, sensations, emotions, etc. Given all the different areas involved in aspects of perception, how is it that we have unitary perceptions: we see, hear, sense, feel all at the same time what is happening during an experience? It used to be thought that all the components of an experience were brought together in a master location in the brain, where consciousness resides, but no such place has been found. It is beginning to seem now that the different areas involved in processing different aspects of an experience become synchronized with each other when experience is unified, creating widespread harmonious electromagnetic patterns. The more unified the consciousness, the more widespread the synchronizations. When I am eating while watching television, thinking about various things, not paying much attention to anything, some subsets of brain areas may be synchronized with each other, but there is no overall harmoniousness. But if in a moment I am fully aware of myself and of everything going on within me and outside of me, the synchronizations may be much more widespread. Different degrees of awareness may correspond to different amounts of synchronization, and perhaps different predominant frequencies. Experience comes together not in space, in a place, but in time, in the present.

Gurdjieff called his school "The Institute for the Harmonious Development of Man." This carefully chosen phrase reflects the purpose of all the aspects of his Work: to bring the organism into a more harmonious overall state of vibration, so that it becomes receptive to the finer vibrations of consciousness. If we believe that the universe emanates from God, and that God's fundamental and highest nature is consciousness, then consciousness is not just an emergent property of a complex brain, but a universal and fundamental energy to which we can be more or less receptive. Just as the rhythm of breathing must be correct for us to assimilate the components of air that we need, Gurdjieff considers that the vibrations of our three brains, or centers, must be appropriately tuned and related to each other so that we can be receptive to this higher energy of consciousness. Consciousness unites our perceptions; just as we metabolize food to produce energy, we can metabolize impressions, transforming them into conscious wholes. Gurdjieff regards the transformation of impressions as our proper highest role in the universal exchange of substances, or energies, because it serves to maintain the consciousness of the universe. In other traditions it is said that God created man so that He could know Himself.

Attention is the main tool to achieve this end. The attention of the mind is brought together with the sensation of the body, permitting a feeling of presence to arise. Meditation, or the sitting exercise, used in many other traditions, is a concentrated form of this practice. Gurdjieff's movements, or sacred dances, require an unfamiliar and intense concentration of attention for their proper execution. And Gurdjieff was an expert at delivering "shocks" to people, bringing them from a state of waking sleep to a sudden awakening of conscience and intense awareness of themselves.

Vibrations serve many purposes in the brain and body, because everything vibrates. Some vibrations have symbolic roles, like speech. The vibrations of movement have a purposive role. Music, however, can be pure vibration, in the sense that music doesn't have to represent anything. It speaks to us directly, particularly to the feeling. It must have a particularly close relationship with the rhythms of the brain, be in tune with them, to have this effect. It speaks to us in a different language, a direct, non-symbolic, one. As such, it can be a force for either good or evil, depending on the kind of harmony or discord it induces in the organism. One of the major components of Gurdjieff's legacy was a large body of music, composed with Thomas de Hartmann, designed to bring us into tune with the highest.

Notes

1. G.I. Gurdjieff, *Life Is Real, Only Then, When I Am* (New York, Triangle Editions, Inc., 1975), p. 144.

2. P.D. Ouspensky, *In Search of the Miraculous: Fragments of an Unknown Teaching* (London, Routledge & Kegan Paul, Ltd., 1950), p. 122.

3. G.I. Gurdjieff, *Views from the Real World* (New York, E.P. Dutton & Co. Inc., 1973), p. 82–3.

THE LIGHT OF THE BEHOLDER

In ancient times, light was regarded as divine, and as essentially synonymous with the highest form of consciousness. When the Egyptian sun god Ra opened his eyes, there was light, and when he shut them, there was darkness. Humanity's task was to become one with the divine light, to develop a corresponding consciousness.

> This then is the message which we have heard of him, and declare unto you, that God is light, and in him is no darkness at all.
>
> If we say that we have fellowship with him, and walk in darkness, we lie, and do not practice the truth:
>
> But if we walk in the light, as he is in the light, we have fellowship one with another. (1 John 1:5-7)

For many, it was possible to "see the light" at least temporarily, and to try to live according to its precepts, but a permanent "enlightenment" was more difficult and rare; those who achieved this were depicted with haloes of light surrounding their heads.

> If the light rises in the Sky of the heart. . . and, in the utterly pure inner man attains the brightness of the sun or of many suns. . . then his heart is nothing but light, his subtle body is light, his material covering is light, his hearing, his sight, his hand, his exterior, his interior, are nothing but light.[1]

Nowadays we tend to regard this ubiquitous equation of light, divinity, and consciousness in the religious and mystical texts of every tradition as apt and appropriate, but only metaphor. For the study of light also gave rise to many of the developments of modern science, and in the process the inner light of the spirit and the outer light of the sun became conceptually separated. The practical results are all around us, from the lights in our homes to lasers, and, via the extended electromagnetic spectrum (of which visible light is one octave of many), to radio and television, cell phones, and so forth. The theoretical results are even more astounding. The mysterious properties of light, which have preoccupied every great scientist from Galileo and Newton to Einstein and Feynman, gave birth to the early twentieth century's twin fundamental theories of modern physics, relativity theory and quantum mechanics, each rigorously verified and yet profoundly paradoxical. Scientists and philosophers continue to this day to struggle with the meaning of these theories, struggle to understand light. Einstein declared, toward the end of his life:

Fifty years of conscious brooding have brought me no closer to the answer to the question, "What are light quanta?" Of course today every rascal thinks he knows the answer, but he is deluding himself.[2]

In his wonderful and meticulously researched book, *Catching the Light: The Entwined Story of Light and Mind*, physicist Arthur Zajonc describes the history of thinking about light through the ages, and how the prevailing ideas about light both reflected and contributed to the general worldview of each age. The "Enlightenment" was a period in the history of Western thought, when science was fresh and full of promise, in which that term meant something quite different than it does today, reflecting the period's enthronement of reason at the top of the hierarchy of human values. Subsequently, reason has taken quite a beating from many quarters, but the paradoxes of modern physics threaten to implode the scientific edifice—the house that reason built—from within, unless a more encompassing view can be found. This view, many are arguing, may bring us back to ancient truths, now on broader foundations, in which inner and outer light are reunited.

What are these ancient truths, nearly left behind by the juggernaut of mechanistic science? Chief among them is that the universe is a purposeful, meaningful, and conscious unity—"all is one"—at the same time that it is almost infinitely differentiated into separate parts. For this unity and diversity to co-exist, somehow each part must contain the whole. In the image of Indra's net, the universe is a web of fine filaments; at each intersection is a jewel which reflects all the others. In this universal organism, human consciousness plays a special role, because of its special capacity to reflect the whole more completely. Humanity was created in God's image, and God created human beings so that He might know Himself.

Were the eye not of the sun,
How could we behold the light?
If God's might and ours were not as one,
How could His work enchant our sight?[3]

In this cosmology, there are hierarchical levels of spirituality, or consciousness—angels and archangels, cherubim and seraphim—and levels of materiality, going in opposite directions. Humanity is unique in its capacity to span most of these levels, but is fallen, and must work for redemption, to turn base metals into gold, to transform materiality into light.

[William of] Conches also drew from Galen the view that food was transformed from matter into a spiritual light in a series of stages. Its first transformation occurred in the liver, where it became "natural virtue."

Passing then on through the heart, it became "spiritual virtue," moving finally into the brain, where it was refined into a luminous wind that animated the organs of sense and provided the interior ray of the eye.[4]

This process of transformation is inextricably entwined with understanding— with perceiving and thereby reflecting back the true nature of the world. In Indian thought, ignorance—*avidya*—plays the same role as original sin in Christianity. (Thus true science and spirituality cannot be opposed to each other.) This understanding is both rational —everything is number, as Pythagoras taught—and beyond rationality, ineffable, undescribable, and non-visualizable.

Its name is unknown:
I simply call it Tao.

—LAO TZU

When we turn to modern physics, the properties of light are strange and mysterious indeed. First relativity, then quantum mechanics, and then the combination of the two have yielded a Pandora's box of bizarre attributes that continue to confound those who wish to add light to the list of puzzles considered solved by science.

We are so used to light that we take it to be part of our ordinary world—the world of people, streets, and houses—but it is not. Light belongs to another, invisible world. For light, like the wind, is invisible in itself; it is only visible by its interaction with things. Deep space is filled with light, but completely dark, except where an object intervenes.

Einstein's special theory of relativity is based on two premises: the consistency of the laws of physics in all smoothly moving (non-accelerating) frames of reference, and the constancy of the speed of light. The first is a familiar property of our world: in a smoothly moving train, a dropped object falls straight down and all other motions are perfectly normal from the point of view of the riders. But the second is not: no matter how fast an observer is moving, light always moves, relative to the observer, at a fixed speed of 300,000 kilometers per second (in vacuum; the speed changes in different materials). One can move faster than sound, as the Concorde does, but it is not possible to catch up with a beam of light. These two premises, paradoxically superimposed, lead to all of the strange results of Einstein's theory: as objects move faster, they become contracted in the direction of motion, become more massive, and time slows down. So nothing with mass can move at the speed of light; if it did, its mass would become infinite, its thickness would become zero, and for it, time would stop. Conversely, particles of light have no rest mass, and so cannot be at rest; they can only move at lightspeed, and for them, there is no time nor space between here and there.

> The speed of light is unvarying. If something comparable existed in spatial terms, it would have the property that no matter where you went, it was always the same distance away from you. You could get neither closer to it, nor farther away from it; the separation would be fixed. This is what infinity is like. No matter how far away from it one travels, it remains the same infinite distance away. As Einstein said, the speed of light plays the role of an infinitely great velocity. Light has no place, but it does have a speed and we are always separated from it by 299,792,458 meters per second.[5]

Another way in which a distance could remain the same no matter where you went is if the distance in question were in a different direction than any you could travel in. One can move as much as one wants on the surface of a sphere, but the distance to the center always remains the same. A two-dimensional being, unaware that the surface was that of a sphere, believing instead that it was an infinite flat plane, and unaware that the distance to the center was of a different nature than distances on the surface, would be very puzzled by the constancy of the radius, if somehow it became able to measure it.

These two analogies suggest very directly that light does not travel in ordinary spacetime, but in a dimension, or dimensions, of its own. For the relation of one dimension to another is as zero to infinity: it takes an infinite number of lines to make a plane, and an infinite number of planes to make a solid. And to travel from one point to another while traversing (for itself) no space nor taking any time, requires that light move in another dimension. Even more intriguing is that although the speed of light, for us, plays the role of an infinitely great velocity, it is not, though huge, actually infinite. This suggests that the dimensions of the world are not perfectly perpendicular, unlike ideal Euclidian dimensions. The earth, for creatures of our size, is *almost* flat, and the speed of light, compared to movements we are familiar with, is *almost* infinite. These deviations from ideal dimensionality may be a key to the possibility of interaction between levels of materiality.

If light has no place, perhaps it has a shape, which would rescue its status as a thing like other things. Here quantum mechanics has removed our footing. The basic finding of quantum mechanics is that light behaves both like a stream of particles, and like a vibrating wave similar to waves on a pond, depending on how one looks. These two aspects are contradictory, but no one has found a way to get rid of either. To make matters worse, the wave, unlike water waves or sound waves, cannot be regarded as a wave in any medium. Scientists have searched in vain for evidence of the "ether," the medium of which light waves are a disturbance. Rather, light is a wave of possibility, which allows one to calculate where a particle of light might be found.

What bears the figure we call light? One thing has become certain, whatever it is, it is not material![6]

So in one sense light is pure form, but not made of anything, and the form depends on the type of observation. In another sense, light, like space, is nothing at all, yet it illuminates everything.

> Throughout history, space has been like the emptiness inside Lao Tsu's clay pot. Ever since we created the concept, space has held whatever we put into it. We have imagined space to be many things, and that act of imagination has had implications for our image of light. Endow space with divinity and light is godlike; discover its shape and light is geometrical; fill it with matter and light is substantial. From Moses to Einstein, the history of light is also the history of space.[7]

Quantum mechanics went further still, discovering that all of the elementary particles of which the world is made, in fact all things, have both wavelike and particle-like properties. The particle-like properties correspond to our illusion of substantiality, but the wavelike properties belie it. But the particles of which things are made—the protons, neutrons, and electrons—differ fundamentally from light particles in that two of them cannot occupy the same space at the same time, so that aggregates of these elementary particles must take up space and become objects. Light particles, on the other hand, do not take up any space, as befits their insubstantiality. In front of my eyes are the light particles from everything I see, as well as innumerable radio and television broadcasts, cell phone conversations, infrared rays, 60-cycle household current vibrations, not to mention the subtle disturbances produced by my own body—electrocardiogram, brain waves, etc. They are all here, needing only a suitable detector to become evident, not getting in each other's way at all. The shape of the tree I see is perfectly preserved, once the lens of my eye has focused the light reflecting from it on my retina, despite the chaos of other vibrations it was mixed with on its way here. And my cell phone picks out my call, unadulterated (usually) by all the others that pass by in the same space.

Still, light is not quite nothing, just as its speed is not quite infinite. Light is affected by gravity, though weakly, and can exert pressure, also weakly.

Whether a scientist sees a particle or a wave depends on what he or she looks for. Quantum mechanics destroyed the materialistic scientist's ideal of complete objectivity, of removing himself from the subject of investigation. Observation—some would say consciousness—plays an inescapable role in determining the apparent nature of phenomena. And certain fundamental features cannot be observed simultaneously: wave or particle can be seen with complete precision in different experiments, but not at the same time. A number of other "complementary" aspects of reality form the very basis of modern physics: where a particle is and how fast it is moving are not precisely determinable simultaneously, nor are its energy and the exact time that it has that energy (this makes it possible for particles to pass through barriers and to appear out of nowhere, if they do it quickly enough). It is as if these complementary attributes were in different dimensions, and the observer's point of view was necessarily single. A spot moving in a circle appears to be moving only up and down if I change my viewpoint by ninety degrees. Limited to one perspective, I cannot see both aspects at the same time.

Another aspect of quantum theory that suggests other-dimensionality is seen in the non-local correlations between elementary particles. Elementary particles, including light, which have interacted at some point become "entangled"; their possibility wavefunctions are combined, so that when they then travel away from each other and are later detected, their properties are correlated. But, in the strange world of quantum logic, one cannot say that the particles had these specific correlated properties during their travels; only on measurement of one of them is a given property defined, and this measurement instantly results in a corresponding property being defined in the distant companion, at faster than lightspeed. Since the predecessors of all the particles in the universe were once interacting, at the beginning of time, or the "big bang", all particles are connected in this way. Thus, in a sense, each part reflects the whole.

There is more, but perhaps that is enough. How we think things are is based on early experience with our bodies and the objects with which our bodies interact. Our brains are molded by this experience, and develop a capacity of visualization that serves us well in navigating through traffic and planning constructions, but it is a limited view of reality. Probing into the fundamental features of the world, we find rules and things that are no longer visualizable. Nor do they obey ordinary logic, which is also based on visualization. Light, the illuminator of all, has not only become invisible, it has become unvisualizable.

However, if we turn our contemplation away from the outer world and to the inner one, as the sages advise, a different reality becomes evident. Like light, consciousness has no place, and no shape. It is invisible yet illuminates everything. It is unimpeded by time or space.

William of Conches' ideas about the transformation of light in the human organism are not so far from modern concepts, for the body is powered by electricity. The energy of food exists in the form of electrical potential energy, provided by the arrangements of the electrons in organic molecules. The energy that creates these configurations is originally drawn from the sun, its light acting in the organic factories of plants. In the body, this energy is transferred to other molecules, first in the intestine and liver, and food is simultaneously degraded to lower-energy wastes. The electrical energy obtained from food provides for all the body's needs, and fuels every cell: a living cell always maintains an electrical potential across its membrane. A dead body is one in which the electrical potentials have dissipated; the lights have quite literally gone out.

The nervous system and brain also depend on electrical energy for their maintenance, and in addition use electrical signals to process information. While the physiological basis of consciousness is not yet understood, recent evidence indicates that it may depend on electromagnetic vibrations—light, though not in the visible range—involving significant portions of the nervous system.[8] The degree of vibratory correlation between different portions of the nervous system may correspond to levels of consciousness: in a relatively automatic mode of functioning, smaller portions of the brain may share enough electromagnetic vibrations to process the information needed for routine thinking and moving, while in a state of greater awareness, there may be a more generalized harmony of vibrations in the brain or even the whole body.

Electrons are held in their atomic orbits by a continual exchange of photons—particles of light—with the nucleus. When atoms or molecules change their energetic state, it is by absorbing or emitting photons. Wherever molecules are vibrating, which is everywhere, light is being produced. The shape of the world comes to us as patterns of light, and is represented in us as patterns of light. Light is a universal medium of attraction and repulsion, transformation and exchange.

If humanity has a special place in the universal metabolism, it cannot be simply to eat organic molecules and produce heat, movement, and waste—the animals can take care of that. And it cannot be simply to build houses and bridges—they will fall down, and in any case only serve our maintenance needs. No, humanity's role must be to *know*, for which we are uniquely endowed. But to know not just practically, so that we can build better bridges, nor even just intellectually, but rather, as the Sufis would say, to know like a lover, so that one becomes one with the cosmos, reflecting it perfectly back to itself; thereby it knows itself more fully.

A lover knows the beloved with mind, feelings, and body. These three ways of knowing are tuned to different aspects of reality. Just as two eyes working together allow the perception of depth, while for one alone the world is flat, when the three modalities of perception work together, the fuller dimensionality of the world is apprehended.[9] Neils Bohr, one of the fathers of quantum mechanics, who pondered long and hard about its implications, thought complementarity to be a general principle. Not only are elementary particle properties—such as position and momentum, energy and time, wavelike and particle-like—complementary, but there is also a complementarity between thought and feeling, and between part and whole.[10] Then, although scientific experiments cannot measure all aspects of reality simultaneously, perhaps the human organism, in its highest states of consciousness, can encompass them. The intellect is specialized for perceiving form, conveyed to us primarily through patterns of light. The emotions are tuned to energy and meaning; the language that speaks to them most directly is music, pure vibration that need not define any shape; in the brain, these vibratory patterns are transformed into electromagnetic ones, light. Color also has a special relationship with feeling. The body, substantial itself, interacts with the manifestations of substance; heat, for instance, also consists of electromagnetic vibrations. All three modalities of perception apprehend light, in its different aspects.

The parts of the Creation, separated as objects, come back together in perception and awareness, in a cosmic circulation. Perceptions are exchanged and shared among us, just as objects are, and, even more than objects, they depend on the human community for their full development. We are as dependent on them for our spiritual life as we are on food for the maintenance of the physical body. Perhaps the universe depends on them as well. From the sun through organic life on earth and through humanity, light undergoes its mysterious transformations.

"That which is called light in creation is, in all its forms and in every being, one and the same spirit, a flame unique." [11]

NOTES

1. Najm Razi, 1256 C.E., quoted in Arthur Zajonc, *Catching the Light: The Entwined Story of Light and Mind* (New York, Bantam Books, 1993), p. ix.

2. Albert Einstein, 1951, *ibid.*, p. ix.

3. J.W. von Goethe, paraphrasing a Neo-Platonic text, *ibid.*, p. 22.

4. Arthur Zajonc, *op. cit.*, p. 29.

5. *Ibid.*, p.269.

6. *Ibid.*, p 123.

7. *Ibid.*, p. 97.

8. There are many references supporting this idea. For one overview, see Gerald M. Edelman and Giulio Tononi, *A Universe of Consciousness: How Matter Becomes Imagination* (New York: Basic Books, 2000).

9. See "Shadows of the Real World" in this volume.

10. Menas Kafatos and Robert Nadeau, *The Conscious Universe: Part and Whole in Modern Physical Theory* (New York: Springer-Verlag, 1990).

11. Paraphrase of Herder, Arthur Zajonc, *op. cit.*, p. 329.

3

THE INNER AND OUTER WORLDS

ONE AND ONE MAKE ONE
The Mystery of the Outer World and the Inner World

Mullah Nasrudin was on a journey, and he stopped for the night in a town where he did not know anyone. He found an inn, and slept comfortably. The next morning on awakening, he discovered to his dismay that he did not know who he was. He thought for a while about his predicament, and then decided to go out into the street to see if anyone might recognize him, and tell him who he was. There were many people in the street, but since he was a stranger in the town, no one recognized him. After wandering around a while, he decided to go into a clothing store: perhaps someone in there...

The shopkeeper pounced on him. "Ah, good sir, I have just the suit for you. Here, try this on." The Mullah complied, and tried on several suits and jackets, none of them quite satisfactory. After humoring the shopkeeper for some time, he turned to him and said:

"Excuse me, my good man, but did you see me come into your store?"

"Well, yes, of course," the shopkeeper replied, puzzled.

"Tell me then," said the Mullah, "how did you know that it was me?"

The greatest riddle, the greatest mystery of all, aside from the Creation itself, is the mystery of the inner world and the outer world, and of their relationship to each other. Perhaps it is even the same as the mystery of the Creation.

This great riddle spews out other riddles, parts of itself, like a volcano. For instance: from the point of view of the outer world, I am nothing, one of six billion little human specks on a tiny planet in a remote corner of a galaxy which is itself one of billions. But from the point of view of my inner world, I am everything; everything that I know, am aware of, is in me: from the stars to subatomic particles, from the ancient Egyptians to my parents and my children; past, present, and possible futures. For me, nothing actually exists outside of my inner world.

That the two worlds are interdependent is obvious: on the one hand, if certain portions of a person's brain are destroyed, all ordinary evidence of an inner world in that person disappears; they become *unconscious*. More limited destruction of parts of the brain leaves gaps in the inner world: language, visual imagery, or memory disappear. On the other hand, a house does not get built without being conceived and planned in someone's imagination. All the materials are there, but the form will not emerge; the materials will scatter with the wind and the rain.

However, the inner and outer worlds exist in different spacetimes. Although the brain appears to be necessary for the existence of an inner world in a given person, no amount of dissection, microscopic examination, computerized imaging, or recording with electrodes will reveal anything that resembles an inner world. One finds structures and activities that are *correlated* with inner world phenomena, but not the inner world itself.

For the localization of outer world events, at least in the ordinary macroscopic world we are familiar with, four parameters suffice: three spatial coordinates and one temporal coordinate. But the inner world cannot be found in outer space at all.

What about in time? Neuroscientist Peter Fox: "The brain exists in space. But now the mind—the mind operates in time alone." [1] Time is a mystery in itself, for from a certain point of view, it does not exist at all: the past is gone, the future is not yet, and the present is an infinitesimal moment between the two. In physics, time simply serves as a marker for the sequence of unfolding phenomena. The inner world requires a bigger time than the outer world; in our inner lives, we do not just move along a steady line from past to future at "one second per second." The future is full of possibilities, only some of which will come to pass outwardly. Even the past has many possibilities, things that could have been but were not. Our minds move around effortlessly in these realms. And the present, especially in a state of heightened awareness, is not just an infinitesimal instant, but a magical fullness. Time manifests its multidimensional splendor only in the inner world.

What is the inner world made of? Some say quantum stuff, since the quantum level has uncanny similarities to the inner world, such as dimensions of possibilities, and the need for observation to choose some possibilities for actualization. Perhaps it is made of patterns of electromagnetic energy, a subset of quantum stuff. But these are also the building materials of the outer world.

Over the centuries, some have argued that only one or the other of these two worlds exists in reality, relegating the other to the status of "epiphenomenon." But the reality of each is undeniable, even though they are incommensurate. Incommensurate but correlated, immiscible but fully intertwined; logically they are mutually exclusive, but paradoxically they co-exist, making reality fundamentally mysterious—accessible, perhaps, only to a transcendent consciousness.

A house cannot exist without building materials, but it also cannot exist without the imagination of its creators. Which is more important?

Both. Which comes first? Although this is an unanswerable chicken-and-egg question, one tends to say: the imagination, the concept, the plan, the vision. These brought about the coalescence of the materials into a form, the house. Similarly, is the inner world primary, the cause of the Creation? "*In the beginning was the Word . . .*" If so, the inner world is not just a private affair. This is another mystery. Even though the inner world of each person is completely private, our consciousnesses feel connected, like the branches of a tree, or like mushrooms linked by underground networks; in a state of full awareness we feel this especially, joined not just by the mediation of external manifestations, but directly. The extreme of this experience is the mystical feeling of being connected to everything. Is this an illusion? Or is the inner world a reality that permeates everything, in which we all partake, more or less, as consciousness permits, just as we all partake of the air, and of the atoms spewed out long ago by supernovae?

What gives the inner world its equal status with the outer is not that there is a chain of neural associations between a sensory input and a motor output; these can be easily accommodated in the outer world. No, the *sine qua non* of the inner world is awareness, consciousness. It is not simply awareness of sense impressions, which is just another way to say that they have been processed to lead to some action—the same thing as the neural chain. The inner world depends on self-consciousness, an awareness of being aware, of being an entity that is aware. An entity that is separate, yet connected to everything.

Since this kind of awareness is intermittent and of variable intensity, interrupted by stretches of more or less automatic reflex actions, and is absent during most of sleep, the inner world of each of us waxes and wanes, comes and goes. Where does it go? Is it always there, only more or less illuminated? After all, when we wake up in the morning, our memories, knowledge, and capacities are generally intact. We are actually speaking about two different aspects of the inner world: its content, and the awareness that illuminates that content. These even seem to be related to different brain regions: the content to the cerebral cortex, and the light to more central brain structures, some of which make up the "reticular activating system," which regulates different states of consciousness.

Actually, there are three elements that make up a state of full awareness: awareness of the outer world, awareness of oneself through inner sensation and feeling, and awareness itself. Each of these involves different brain regions, and it may be that coordinated electrical activity between separate parts of the brain underlies the sense of self-consciousness. If so, the physical correlate of an inner life is a sufficiently complex electromagnetic pattern at the level of the entire nervous system.

For an entity to have an inner life, it must have a sensation of itself, of being an entity. But no entity exists in isolation in an outer sense—it needs food, air, and other materials to maintain itself. Similarly, no entity is isolated in an inner sense: a large part of its inner life consists of impressions of the outside world. Outwardly and inwardly, an entity is like a living cell, with a semipermeable membrane that both defines it and connects it with the outside, allowing some substances to pass through in each direction and blocking others, in a dynamic equilibrium. This equilibrium both maintains the cell as a semi-independent entity, and provides its interaction with the world at large. Hence, by analogy, the reciprocal relationship between self-awareness and the sense of being connected to the outside. One of the crucial functions of a cell membrane is the establishment of an electrical potential across itself, which it accomplishes by differential permeability to different ions in the extracellular and intracellular fluids. Similarly, perhaps, a sense of oneself aware of the world requires the establishment of an electromagnetic entity within the body, by the interaction of large portions of the brain. This then serves as the semipermeable membrane of the inner life, by which we attend to some things and not others, build percepts and concepts from subsets of elemental sensations, choose some actions over others, and communicate with the outer world.

If this analysis is correct, other entities that have complex energetic patterns, that are maintained in dynamic equilibrium and interaction with the rest of the world, may have an inner life. The solar system, for instance, consists not just of rotating and revolving spheres held together by gravity, but also of complex electromagnetic interactions between the sun and the planets, which have been increasingly discovered in recent decades—the solar wind, the magnetosphere of the earth, and the earth's magnetic tail which reaches as far as the moon's orbit, intersecting with the moon when it is full. Is this the conduit by which our emotions feed the moon? (According to Gurdjieff, the moon needs this food to maintain itself and to grow.)

Are consciousness and matter the inner and outer aspects of the same thing? Early in the twentieth century, Einstein discovered the interchangeability of matter and energy. This was related to a change in the concept of time, which was needed to accommodate the bizarre consequences of relativity theory, such as the invariability of the speed of light regardless of the speed of its source, and the relativity of simultaneity. Matter and energy are thus two aspects of the same underlying reality. The third aspect, then, is consciousness. This is not a generally accepted notion, but hints exist in quantum theory, which requires a dimension of possibilities—a further enlargement of time—and makes observation a fundamental component of reality.

The foregoing might suggest that all semi-independent entities, from subatomic particles to galaxies, partake in some way in consciousness, and have some kind of inner life. Some believe this, but no one would attribute to molecules or cells the kind of awareness we have. Is it a matter of degree, or is there a threshold that makes possible the kind of inner world we are privileged to inhabit? One view is that it is the development in us of the capacity for abstract and symbolic thought, that in harmonious combination with sensations and emotions, makes it possible for us to resonate with the full reality of the universe. Thought by itself easily slips away from reality into fantasy, but when it is blended with sensation and feeling, we can enter into the magical present moment, no longer infinitesimal, but all-encompassing. Each of these capacities is tuned to a different dimension of time and to a different aspect of the world: sensation is allied with matter, and thus with the flow of ordinary time; thought is at home in the realm of all possibilities; and feeling, linked with energy, mysteriously connects us with eternity.

Yes, the inner world is very elusive (so is the outer world—ask any physicist). Who am I, then? Am I the contents of my skin, this sac of organs, tubes, and fluids? Or am I the contents of my inner world, the totality of all I am aware of? Does this "I" expand and contract, come and go, with the fluctuations of my consciousness, moment by moment? Or am I everything I have ever been aware of, even those things I have forgotten? Will I disappear when my body disintegrates, or is my inner world, or some aspect of it, supportable by some other structure? There are enough mysteries here to occupy armies of philosophers, theologians, mystics, and scientists for thousands of years, and they have indeed been so occupied. But the riddle remains, answerable only to other riddles.

> *A man had the opportunity to make a long voyage to see the Buddha. Knowing that the Buddha was very busy and had only a short time to see him, and that the opportunity would not repeat itself, he thought long and hard to find the best way to get answers to the basic questions of life which had preoccupied him for many years. When the audience came, he asked the Buddha:*
>
> *"Holy Master, please tell me, what is the most important question of all, and what is the answer to it?"*
>
> *The Buddha replied: "My son, the most important question is the one you have just asked, and the answer is the one I am giving you now."*[2]

NOTES

1. G. Montgomery, "The Mind in Motion," *Discover*, March 1989, pp.58–68.

2. The tales at the beginning and end of this article were told to me by Robert Brown.

Shadows of the Real World

If therefore thine eye be single, thy whole body shall be full of light.
(Matthew 6:22)

How is it, that seeing only shadows, we are capable of perceiving reality?

In Plato's allegory of the cave, he suggests that our ordinary perception shows us only shadows of the much more real, ideal world of immutable forms, the world of being, not becoming. Many philosophers, mystics, and others have held a similar belief: that there is an enlightened view, resulting from a higher state of consciousness, when the "doors of perception have been cleansed," compared with which the world that we ordinarily perceive is like a world of shadows. The descriptions vary, and there is always the disclaimer that words cannot adequately describe this perception, since words belong to the shadow world.

> *It cannot be called void or not void,*
> *Or both or neither;*
> *But in order to point it out,*
> *It is called "the Void."*[1]

Nonetheless, from the descriptions there emerges an unmistakable sense that the seers are all seeing the same thing. How is it possible that with our senses and our incredibly intricate nervous system we ordinarily perceive only the shadow world, and yet somehow are capable of passing to this other level of perception?

A shadow is a two-dimensional projection of the outline of a three-dimensional object. Some information has been lost, some is retained: the outline is there, at least from one perspective, and the shadow's movements follow those of its object. A given three-dimensional object can cast any number of different shadows, depending on the location of the source of light. Given a sufficient number of shadows, one can, using mathematical reasoning, reconstruct a very good approximation of the solid object. This process is used, essentially, in deducing the structure of molecules from x-ray diffraction patterns. Generally, much of the scientific elucidation of the workings of nature is based on an analogous kind of reconstruction: many partial views of a phenomenon, derived from experimental observations, are synthesized into a coherent picture.

In these instances, the synthesis of a more complete view of reality is done by the mind, by reasoning. In Plato's view also, reasoning, particularly mathematical reasoning, seems to have the primary role in perceiving the world above the shadows. Yet in other traditions, such as Zen Buddhism, reasoning and mental activity in general, at least ordinary mental activity, are regarded as the source of illusion, impediments to the direct perception of reality.

> No thought, no reflection, no analysis,
> No cultivation, no intention;
> Let it settle itself.[2]

Regardless of this apparent contradiction, the perception of reality which is the attribute of a higher state of consciousness is always described as a direct experience, not deduced like the logical reconstruction of an object from its shadows. Even in Plato's allegory it is the direct perception of the world of the sun which is described.

Our visual system constantly uses shadows, or shading, to deduce the shapes of objects (Figure 1), rapidly and effortlessly performing complex neuronal calculations which are unseen by our awareness, so that what is presented to consciousness is a direct, immediate perception.

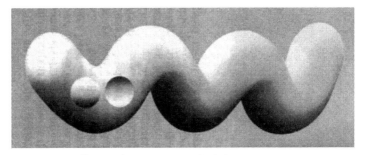

FIGURE 1 • *By permission of Oxford University Press*

An analogous perceptual feat is depth perception. How can we directly perceive three spatial dimensions with two eyes, when each sees only a two-dimensional view? This example is useful, because much is understood about the mechanisms of depth perception, and because the difference between two-dimensional vision and three-dimensional vision is similar to the difference between our ordinary perception and another level of perception we sometimes experience, in the direction of the direct perception of a higher reality that Plato and others allude to.

Close one eye and observe what you see. At first the world may not look any different. Then you may notice a curious flatness, curious not so much because it is there—after all, you probably expected to lose your depth perception—but because it is so subtle. It may not even be obvious that anything has happened to the third dimension, because using a variety of perspective cues, you still know how far away things are and that one thing is in front of another. You can even reach for objects quite accurately. But in fact the world has the visual quality of a painting or a photograph. After a while you may forget that your one-eyed world is in any way deficient. But now open the other eye; the return of the direct experience of depth is striking. There is a fullness, a solidity and yet a roominess—a reality—to the world that was not there a moment ago.

Many people function perfectly well with only one eye, and others have eyes which are misaligned so that they cannot use the two together. They have never had the direct experience of visual depth. One can readily imagine how difficult it would be to describe it to them. One would have to resort to analogies: "It's like the difference between seeing something and also being able to touch it, or between black-and-white and color," or to evasion: "It really looks the same, yet it's completely changed," or finally: "Well, I really can't describe it; it has to be experienced."

Now look out at the world, or listen to it with your eyes closed. At the same time be aware of your body, so that you feel: "I am looking," or "I am listening."

This effort of double awareness has of course been described many times before. It is described very precisely in the Buddhist *Sattipatana Sutta*. More recently, P.D. Ouspensky, attempting to understand what G.I. Gurdjieff meant by "self-remembering," formulated it as follows: "When...I try to remember myself, my attention is directed both towards the object observed and towards myself....Having defined this I saw that the problem consisted in directing attention on oneself without weakening or obliterating the attention directed on something else. Moreover this 'something else' could as well be within me as outside me." [3]

When I try this experiment, the first thing I notice is how easily my attention repeatedly wanders off into thought and fantasy. But if I am able to maintain this double awareness, the world acquires a solidity, a fullness, a quality of reality, of being alive, that it did not have before. This is as subtle and as real as the difference between one-eyed and two-eyed vision. By comparison, my ordinary perception is lifeless, not even really perception, but more like an intermittent trigger for various flights of fancy.

This new perception is not actually new. I have experienced it before, during intense or striking moments. My vivid memories are of moments when this occurred. The dividing line between "ordinary" perception and this enhanced awareness is not so clear, because dimly and for very brief moments it probably occurs many times a day, and throws its light on the moments just before and after. At one extreme, we have all had the experience of "waking up" after several hours of our usual activities, or after driving many miles, and realizing we simply weren't there for all that time; at the other, efforts at self-remembering, life events, or just random good luck can lead to a sudden and dramatic change in perception, which has the quality of another dimension.

The higher dimensionality of this perception manifests in different ways. I may become aware of the constant motion of everything. Everything was moving before, and I was able to catch a ball and not be hit by a passing car, but now the experience seems direct: from moment to moment everything is always changing. By contrast, in my usual state I take snapshots, label them, and file them away, so that I can get on with my dreams. Another aspect of the experience is that I, the perceiver, become solid and real, yet transparent and light; at the same time both my existence as a separate entity and my interconnectedness with the world are directly experienced—the world and I interpenetrate one another. I may truly *see* that another person is like myself, a living, conscious being who has an inner life as I do; only under these conditions is it possible to love one's neighbor. I may become directly aware of where I have been just a moment ago, and even of where I will be a moment from now, as if I were suddenly lifted above the winding road I am traveling and could see its layout. The present—which in my usual state is a nonexistent moment (between a nonexistent past and a nonexistent future)—becomes tangible.

In order to further explore this analogy, it is necessary to understand something about the mechanism of depth perception. The visual image projected by the optics of the eye onto the retina in the back of the eye, which registers the image, is necessarily two-dimensional, like a photograph, and so is the partially processed image sent on from the retina to the visual part of the cerebral cortex in the back of the brain. In order to perceive depth, the visual cortex has to compare the two images from the two eyes, which are slightly different because each eye sees the world from a slightly different angle. Thus, a lamp shade, seen from above, might look like A to one eye and B to the other. (Figure 2)

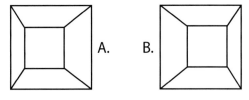

A.   B.

FIGURE 2

If you can cross your eyes just enough to produce a superimposed image of the two drawings above, you will see a lampshade in three dimensions.

In the visual cortex, cells receive converging inputs from the two eyes for each location in the visual scene, and by detecting the slight displacements of corresponding points in the two images, the depth of each point is determined, and a vivid, three-dimensional percept is synthesized. Interestingly, the sense of depth can be produced by this mechanism even though there is no actual object in each eye's image. In Figure 3, a random array of black and white pixels is presented to each eye, the same array for each, except for a central square in which the corresponding pixels are offset horizontally. By fusing the two images (in the same manner as for Figure 2) the central square is seen vividly displaced in depth (it can take a number of seconds for this percept to emerge). This phenomenon, discovered by Julesz in the 1960s (although some earlier examples have been found), provides an insight into the workings of the cerebral cortex. Without a shape in each eye's image, how does the visual cortex know which pixels should

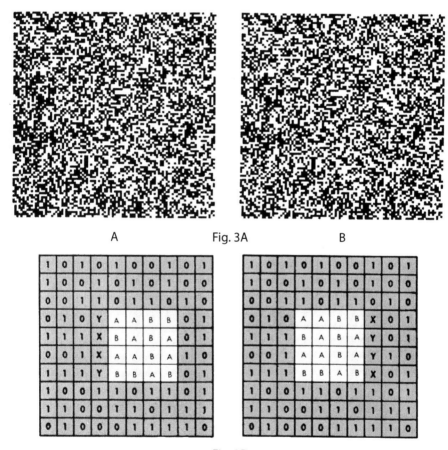

A   Fig. 3A   B

Fig. 3B

FIGURE 3 • B. Illustration by a small array of the way Fig. 3A has been generated.
Courtesy Alcatel-Lucent USA

correspond to each other? In the absence of further constraints, any pixel of one color in one eye could be paired with any other nearby pixel of the same color in the other eye, resulting in a hodgepodge of perceived depths. Since the visual system sees a central square uniformly at one depth, it must have a way to decide on a consistent disparity which most economically describes a consistent surface at one depth. Since a given cortical cell is only directly analyzing a tiny portion of the array, all the cells responding to the overall pattern must somehow communicate with each other to arrive at the most economical solution.

The cerebral cortex is probably the anatomical substrate for the *contents* of our ordinary consciousness. Different areas are devoted to vision, the other senses, language, the organization of intended movements, and more abstract thinking and planning. Yet one of the intriguing aspects of the cortex is its uniformity: despite the widely different functions served by different areas, the general organization at a cellular level is strikingly uniform, although there are some differences between one area and another. So the cortex can be regarded as an organ, like the lung or kidney, which takes in certain substances, has them interact in specific ways, and puts out transformed substances— oxygenated blood and exhaled air in the case of the lung. In the case of the cerebral cortex, in each area nerve fiber tracts bring in information of certain kinds, and these combine in such a way that a transformation is effected, and the processed information is relayed to other places—other parts of the cortex or nervous system. The exact nature of these transformations is not understood, but the example of stereoscopic vision gives a sense of it. Visual information from each eye is brought to the visual cortex, and the disparities of points in each little portion of the visual field are computed in parallel for the entire visual field. These disparities are all correlated with each other to arrive at a realistic computation of the depth of various objects.

Other parts of the visual cortex perform similar operations: an area devoted to color vision correlates the colors in the visual field to arrive at color constancy, so that colors appear the same to us under widely varying illuminations. We do not notice the orange tint that covers everything in incandescent light, as daylight film would, because the orange tint, being a uniform bias, is filtered out by this part of the visual system. Other parts of the cortex, insofar as they are understood, seem to function similarly: parallel point-by-point analyses coupled with massive cross-correlation, over a certain perceptual space. The same kind of process allows us to see the shape of things from the shadows that are cast, as in Figure 1. The visual system assigns a direction for the source of light which fits the overall pattern of shadows, and the two circular areas are then seen as convex and concave, respectively.

These computations yield perceptions which have the quality of being direct, immediate, and real. We are aware of the perceptions, but not the elaborate computations performed by the cortex that make them possible. On the other hand, when we analyze something mathematically, or when we deduce something logically, it does not have the quality of a direct perception. Yet similar sorts of calculations are going on. What is the difference? One difference clearly is that the cortical computations that result in perceptions are hidden; only the result is seen by our awareness. Another, related difference, has to do with time: in direct perception, all the computations are performed within a short amount of time, so as to appear instantaneous—they occur within the perceptual present moment—whereas when we analyze something logically it takes a longer time.

How long is the present moment? If time is regarded as a line stretching from the past through the present to the future, the present is an infinitesimal instant. However, from the point of view of our perceptual experience, the present moment must have a certain, perhaps variable, duration. Various phenomena indicate a general range. Prior to a voluntary movement, a buildup of electrical potential occurs in the premotor cortex starting about a second before the actual movement. Direct cortical stimulation with electrodes, which can be performed during brain surgery, can produce conscious sensations corresponding to the area of cortex stimulated, but a conscious sensation takes about a half second to appear, and a similar—perhaps somewhat shorter—delay probably occurs between the arrival of naturally evoked sensory or perceptual information at the cortex and its conscious appreciation.[4] The shortest reaction time is about a fifth of a second, and in such situations we probably react before we are aware of it. Reading, we move our eyes jerkily from one spot on the page to another several times a second, and looking around at things we move them with a similar frequency. Short term memory, for lists of numbers or words, can hold about seven items, which can be presented at variable speeds, but typically over a few seconds. Listen carefully to someone speak. The individual phonemic sounds can be heard in sequence, but the meaning appears in layers with words, then phrases, then larger segments. The individual words can be replayed precisely in the mind only slightly longer than it takes for them to be assembled into meaningful phrases; then the individual phrases are kept in a sort of shorthand only until larger meaningful units are assembled. So, from these disparate examples, one can infer that the perceptual present moment is roughly one to several seconds long.

Our nervous systems are always occupied with many different things simultaneously. We can all walk, chew gum, think or daydream, and do many other things at the same time. Many, if not most, of these activities go on automatically, without attention or awareness, and it seems efficient that we not be consciously concerned about navigating irregularities in the sidewalk or avoiding passersby while occupied with serious window-shopping. Our attention is usually only on a small portion of our cerebral activity, and typically jumps about from one thing to another. Our awareness is fragmented. By contrast, when the effort of self-remembering described above is made, our attention is more widely distributed, illuminating more of what is happening in the present moment. As I see, hear, and sense what is around me, I am also aware of myself, my posture, movement, inner sensation. I am aware of these things simultaneously, not sequentially. Everything in the present moment is brought into synchrony, into correspondence, so that the possibility of making a broader cross-correlation, a further perceptual transformation, appears—just as the images from the two eyes must be in spatial register for depth perception to be possible. This probably corresponds to the heightened perception of reality that is experienced.

G.I. Gurdjieff placed great emphasis on the idea that man has three brains, which he called intellectual, emotional, and instinctive-moving. This division has been made by many others, and corresponds in a direct way with our experience, but Gurdjieff elaborated on this idea in a unique way, regarding genuine consciousness as possible only when the three brains, or centers as he sometimes called them, are brought into correspondence in a way which ordinarily only occurs for brief moments.

> We must understand that every normal psychic function is a means or an instrument of knowledge. With the help of the mind we see one aspect of things and events, with the help of emotions another aspect, with the help of sensations a third aspect. The most complete knowledge of a given subject possible for us can only be obtained if we examine it simultaneously with our mind, feelings, and sensations. . . . In ordinary conditions man sees the world through a crooked, uneven window.[5]

> Our mind has no critical faculty in itself, no consciousness, nothing. And all the other centers are the same. What then is our consciousness, our memory, our critical faculty? It's very simple. It is when one center specially watches another, when it sees and feels what is going on there, and, seeing it, records it all within itself.[6]

How does a change of perception come about? The three brains, or three parts of the nervous system, are massively interconnected, even interleaved. This is true in our ordinary state; no higher consciousness is necessary. So what changes in another state of consciousness? Experientially, the change is both a focusing and a broadening of attention, so that more of the elements of the present moment are brought together. Ordinarily, the brain is occupied with half-received impressions and at the same time with daydreams, memories, expectations or fears for the future, etc. The same portions of the visual cortex which serve sight also serve visual memory and imagery; the inner screen is usually covered with several images simultaneously, so that none is vividly clear. Looking at the whole nervous system, there is a veritable chaos of relatively random associations. In a state of presence, the activities of the nervous system, of all three brains, are brought into correspondence with what is happening *now*. How this results in a unitary conscious experience is the great mystery of neuroscience; no scientific answer is available at this time. But I, like some others, think the direction of inquiry lies in looking at certain aspects of the brain's electrical activity.

Nerve cells communicate by electrical signals, which travel along the axonal processes that connect one cell to another at a maximum speed of about 100 meters per second. Further delays occur in the transmission of a signal from one cell to another. It takes a minimum of 100 milliseconds for a visual signal to evoke an eye movement, and as mentioned above, it may take a significant fraction of a second for a cortical pattern of activity to impress itself upon consciousness.

But another aspect of the brain's electrical activity—one of the most striking and least understood aspects—is its rhythms: the brain waves, produced by large groups of cells that are synchronously and rhythmically active. The predominant rhythms change with changes in state, being slower in deep sleep (a few cycles per second), and more rapid in alertness. The alpha rhythm, about ten cycles per second, predominates in the relaxed awake state with eyes closed. Faster rhythms, averaging around forty cycles per second, have recently been thought (though this is controversial) to play a role in binding percepts together. Since any sensory perception requires many parts of the cortex to be active in deciphering its various aspects, and since there does not seem to be any one downstream place where all of these aspects come together, how do unitary percepts arise? The prototypical experiment showed that cells in two separate parts of the visual cortex which were responding to two parts of an elongated line segment had synchronized electrical activity, whereas if the cells were equally well stimulated by two separate line segments, their activities were not so synchronized.[7] Further experiments have extended this concept, of the relevance of synchronized rhythmic activity between different parts of the nervous system. For example, cats trained to simply watch one visual stimulus, and to respond to another by pressing a lever, showed a greater synchronization between the areas of cortex devoted to vision and to sensation and movement during the second condition than the first.[8]

Meditation seems to result in more synchronized brain wave activity.[9] So, generally, synchronization of brain rhythms may be an important feature underlying all aspects of consciousness, from simple perception to higher states of awareness. Locally, synchronization may bind percepts together; globally, it may give rise to the sense of self, of a unitary consciousness which includes both the sensation of the physical body and all that is being perceived from outside. This is not to imply that the brain strives to all oscillate at the same rhythm: the information content of conscious perception would require multiple frequencies, but an important role may be played by the *harmonization* of the activities of different parts of the nervous system. The inner life, not locatable in space, may be "located" in time (possibly multidimensional), being in essence a complex electromagnetic harmony.

There are clearly degrees of consciousness, from our ordinary disjointed half-sleep, in which most of what occurs is not conscious at all, to the state of self-awareness described above, to states in which there is an even more expanded perception, so that what is seen in a moment would take hours to explain, as described by scientists or mathematicians when there is a flash of insight about a complex problem. Mozart could hear an entire composition all at once.[10] And Gurdjieff said: "*Consciousness* is a state in which a man *knows all at once* everything that he in general knows and in which he can see how little he does know and how many contradictions there are in what he knows."[11]

In emergencies, people may report an expansion of time, so that everything is experienced with great clarity in slow motion. In mystical experiences, what is seen in an instant is so full of content and meaning that it defies sequential description.

The harmonization of electromagnetic rhythms which may underlie these changes in the state of consciousness might make possible an entirely different level of functioning of the nervous system, in which the mechanisms of perception are no longer confined by ordinary neuronal signaling, limited to meters per second, but function in part at the speed of electromagnetic wave propagation, at light speed. The "light of consciousness" may be more than a metaphor.

Notes

1. Madyamika Shastra, XV.3, quoted in A.W. Watts, *The Way of Zen* (New York: Mentor Books, 1959), p.71.

2. *Ibid.*, p. 85.

3. P.D. Ouspensky, *In Search of the Miraculous: Fragments of an Unknown Teaching* (New York: Harcourt, Brace & World, Inc., 1949), p.119.

4. B. Libet, "Cerebral processes that distinguish conscious experience from unconscious mental functions," in *The Principles of Design and Operation of the Brain*, J.C. Eccles and O. Creutzfeldt, eds. (New York: Springer-Verlag, 1990), pp. 185–200.

5. Ouspensky, pp. 107–108.

6. G.I. Gurdjieff, *Views from the Real World* (New York: E.P. Dutton & Co., Inc., 1973), p. 271.

7. W. Singer, "Self-organization of cognitive structures," in Eccles and Creutzfeldt, pp. 119–128.

8. C. Chiang, A. von Stein, P. Konig, "Synchronous activity between primary visual and sensorimotor cortex in the awake behaving cat," in Abstracts of the Society for Neuroscience, 1996 Annual Meeting, p. 643.

9. M. Murphy and S. Donovan, *The Physical and Psychological Effects of Meditation. A Review of Contemporary Meditation Research with Comprehensive Bibliography 1931-1988* (San Rafael, Calif.: Esalen Institute Study of Exceptional Functioning, 1988).

10. R. Penrose, *The Emperor's New Mind* (Oxford: Oxford University Press, 1989), p. 423.

11. Ouspensky, p.155.

IMAGINATION

According to both G.I. Gurdjieff and neuroscientist Paul MacLean (although perhaps in very different senses), a human being has three brains. In Gurdjieff's terminology, these are an instinctive/moving brain, an emotional brain, and a thinking brain. One of the main functions of the third brain—the one whose size and power in us distinguishes us from all other animals—perhaps its main function, can be called imagination. Imagination is the ability to make and manipulate images in one's mind. Pragmatically, these are images of external reality, whose purpose is to enable navigation and action in the world, ultimately to ensure survival and reproduction. We don't bump into things much, and can plan our meals well ahead, as well as fantasize endlessly about the opposite sex, which sometimes leads to action. In the view of many scientists, this is the origin and purpose of imagination. However, these are not its limits: the manipulation of images in human minds undergoes many levels of abstraction, evidently many more than in the minds of other mammals. A particular chair invokes the concept of "chair," widely connected to other concepts: furniture, comfort, posture, gravity, space, materials, molecules, atoms, elementary particles, and the mathematical abstractions that—strangely and magically—characterize those particles and their interactions. Oddly enough, the "imaginary numbers," derived from the square root of minus one, are required to mathematically represent elementary particles and their transformations.

The outer world is the world of manifestation, the inner world that of imagination. Yet in many respects it is the inner world that makes possible the outer. We do not just reflect the outer world, we create it. In many circumstances we order it—prevent its fall into disorganization, driven by entropy. Yet nature is also highly ordered, by what are ultimately mathematically inevitable patterns and relationships. If a God created the world it is through these laws; perceiving and manipulating them, we resemble the creator.

There are experimental studies that suggest that we do not become consciously aware of a sensory stimulus until about half a second after its onset.[1] Since many of our reactions occur more quickly than this, it can be argued that our conscious choices are illusory, after-the-fact rationalizations, and in many circumstances this can be clearly shown. So "free will" does not exist. But this does not take into account the power of imagination. We can anticipate various possible reactions to a stimulus, and prepare the brain to react in one way rather than another. The sudden appearance of a visual stimulus in the periphery of the field of vision usually elicits an eye movement toward the stimulus, to assess its significance. A subject can be instructed, however, to look away from the stimulus when it appears, and will do so quickly and reliably; young children and people with dementia have difficulty with this task.

One can observe in oneself a constant interplay between perception of and reaction to external phenomena, and imaginative anticipation and prediction. This occurs on many levels: catching a ball is such an interplay, as is building a bridge. Listening to another person speak, we are on the one hand building meaning as syllables become words, words phrases, phrases sentences, and sentences paragraphs, while at the same time anticipating what will come next, so that a grammatical error or unanticipated end of a sentence causes surprise. The cumulative meaning falls into a partially predetermined framework. After spending some weeks in Spain and then going to France, I found myself hearing Spanish phrases being spoken in the street by passersby, although they were in fact speaking French, and even though I knew French much better than Spanish. As we go into a room and look around, what we see is largely dependent on a preconceived template of a room, based on long experience, especially in childhood. Persons born blind whose sight is restored as adults are notoriously unable to make sense of their visual impressions.

We are constantly predicting the immediate future, mostly unconscious of its mysterious unpredictability, and, being afraid of the unknown, our minds often quickly incorporate unexpectednesses into a seamless known fabric: I knew all along that's what would happen.

In fact, the future is both largely predictable and completely unpredictable, but we do not live with this paradox, because for the most part we do not live consciously in the present. What is the present moment? Is it my awareness of sensory stimuli? But these actually occurred in the recent past; it took some time for me to be aware of them. Is it my awareness of what I am going to say or do next, already formulated but not manifest until the near future? Sensory reception and mental anticipation thus frame the present. In Gurdjieff's movements, or sacred dances, one is asked to maintain a constant awareness of bodily sensation and at the same time to visualize the next position to be taken. Thus the present comes into existence. But most of the time our thoughts are wandering in the distant past and many possible distant futures and we are unaware of the present. Our imaginations are so powerful that, unbridled, they easily detach themselves from the body and from the present, and our bodies are left to muddle through on their own.

The problem of time is a very profound one. Einstein's relativity theory demonstrates the relativity of simultaneity, so that two observers moving relative to each other can perceive different time relationships between two events: for instance one says that event A occurred before event B, and the other the opposite. This is similar to the fact that the spatial distance between two points can appear different depending on the angle of view. These considerations make time a fourth dimension, and four-dimensional reality a fixed framework. Our awareness moving along the time dimension gives us the illusion of events unfolding, but those events are already there in reality. In such a scheme there is obviously no room for free will.

The distinction between the past, present and future is only a stubbornly persistent illusion.

—ALBERT EINSTEIN[2]

Time present and time past
Are both perhaps present in time future,
And time future contained in time past.
If all time is eternally present
All time is unredeemable.

—T.S. ELIOT[3]

But quantum theory gives us a different picture, one in which many possible outcomes seem to exist for an event prior to its measurement. Quantum theory and relativity theory have so far resisted being completely reconciled with each other, although each has proved to be incredibly accurate in describing experimental outcomes within its purview. It is possible that a future theory that brings these two, each already supremely paradoxical, theories together will also help to elucidate the nature of time, consciousness, and free will.

Even Einstein confessed, near the end of his days, that the problem of the now "worried him seriously." In conversation with the philosopher Rudolf Carnap he conceded that there is "something essential about the now," but expressed the belief that, whatever it was, it lay "just outside the realm of science." Maybe, maybe not.[4]

The mind has qualities that resemble phenomena at a quantum level. The simplest examples come from ambiguous figures, such as the Necker cube, or the faces and vase illusion, that can be seen in one of two ways, but not both simultaneously.[5] The visual object is neither one thing nor the other; the mind makes a choice, and the object becomes that in the mind. We are constantly making such choices, as when I heard Spanish in the French streets; our reality is to a significant degree a construction of the mind. And without this capacity of imagination, we could not construct the externally real objects that have transformed the earth.

The perceived present must have a certain duration, due to the delays in perception and manifestation, that straddles the moment in time in which

I am now physically present. Similarly, the essence of a pendulum cannot be captured in an instant, but depends on its cycle. In this cycle energy is constantly being transferred back and forth from one form to another, from the potential energy of gravity to the kinetic energy of motion; that is what maintains the cycle. All vibratory motion is of this nature, and the world is essentially made of vibrations. One can speculate that the constant interplay between the immediate past and the immediate future that occurs when the present is consciously attended is also a vibration, serving a mysterious role in the self awareness of the universe, and made possible by the proper use of imagination. Thus, perhaps, is time redeemed.

NOTES

1. Libet B, Pearl DK, Morledge DE, Gleason CA, Hosobuchi Y, Barbaro NM: Control of the transition from sensory detection to sensory awareness in man by the duration of a thalamic stimulus. The cerebral time-on factor. Brain 1991; 114:1731-1757.

2. www.brainyquote.com/quotes/authors/a/albert_einstein.html

3. T.S. Eliot, *Four Quartets*, "Burnt Norton."

4. Paul Davies, *About Time: Einstein's Unfinished Revolution* (Simon and Schuster, New York, 1995), p. 77.

5. For a picture of the Necker cube, see "The Cosmic Metabolism of Form" in this volume.

AWAKENING THE EMOTIONS
The Transformation of the Emotional Life

Why do people, the world over and since time immemorial, fast, abstain from sex, and put themselves in deliberate danger or pain? Many initiatory rites involve such practices. They are used as a help to attain communion with God, or the spirit world, or to find a feeling of oneness with the world which is free of the prison of ordinary egoistic concerns.

Although evolutionary theorists can and do explain away all of our human characteristics by the imperatives of species survival, these behaviors appear counterproductive. For hunger, sexual desire, fear, and avoidance of pain are deeply ingrained, hard-wired if you will, instinctive-emotional drives that protect the individual from death and the species from extinction. Without these and other basic tools of self-preservation, nature's great cornucopia of creatures would not exist.

From this point of view, what are our emotions? They are one aspect of neural systems which mobilize our bodies and minds to deal with survival necessities. When my body sends signals that it requires nutrition, food occupies my thoughts, becomes desirable, even beautiful, and my nervous system is driven to organize itself and me to obtain some. Knowledge and habits on the subject are brought to the surface; my body is readied to move toward food. Sexual hunger functions in a similar way. The nervous system is organized as a series of hierarchies, so that simple reflexes in the spinal cord are organized into coordinated patterns of movement, such as stepping, by neuron pools in the spinal cord itself, and then into more and more refined and complex movement patterns with sensory feedback at progressively higher levels of the nervous system. Above the level of complex coordinated movement are the instinctive drives, which coordinate the entire organism.

The neural substrate for fear, the response to danger, is one of the best studied. In the 1920s, Walter Cannon characterized what he called the "fight or flight" response: blood pressure rises, heart rate increases, blood is shunted away from the digestive system and to the muscles, the pupils dilate, and other phenomena mediated largely by the sympathetic nervous system occur, which prepare the body for the emergency. These responses are coordinated by structures deep in the brain, such as the hypothalamus and amygdala, which function below the level of consciousness. More recent research, particularly by Joseph LeDoux, has shown that learned emotional associations can also be formed at this level.[1] Thus phobias, panic attacks, and more everyday kinds of reactions and anxieties are both learned and then activated largely subconsciously, though they also interact with consciousness. The emotional brain has its own memory, distinct from conscious memory. So, when an emotional reaction occurs, the reasons the thinking brain comes up with for its occurrence may or may not reflect its origin. Self-study shows this to be so: I am attracted to this person, dislike that one, am afraid of mice, find pleasure in listening to rain—often not knowing why, or finding that the explanations my mind produces have a suspicious flavor of after-the-fact confabulations.

Calling instinctive drives "emotions" may not seem quite right. For many psychologists and neuroscientists, emotions are rather the subjective manifestations, the inner conscious accompaniments, of instinctive drives. William James, father of modern American psychology, lists under human instincts not only sucking, clasping objects, carrying them to the mouth, locomotion, and vocalization, but also emulation, rivalry, fear (of various kinds, e.g. of strange men and animals, dark places, and solitude), acquisitiveness, curiosity, play, constructiveness, secretiveness, sociability, modesty, jealousy, and love. James considered emotions to be the subjective aspects of these instinctive impulses, and he postulated that the emotions, so defined, were derived solely from sensory feedback from the body. So fear, either of a completely innate instinctive sort, or derived from such by an experienced association, gives rise to the variety of bodily changes described by Cannon, which, being sensed by consciousness, in turn produce the subjective emotion. Though "common-sense says, we lose our fortune, are sorry and weep; we meet a bear, are frightened and run; we are insulted by a rival, are angry and strike. . . . the more rational statement is that we feel sorry because we cry, angry because we strike, afraid because we tremble."[2] This hypothesis continues to be debated today. Certainly, it is difficult to imagine the full experience of various emotions without the inner sensations that accompany them. On the other hand, it is argued that emotions can arise much more quickly than their bodily manifestations. LeDoux believes that in some instances there may be an "as-if" feedback, whereby a brain representation of (previously experienced) bodily sensation can substitute for the sensation itself. Having experienced the sensations associated with fear, a memory of these sensations could be quickly activated, followed by the actual sensations themselves.

Antonio Damasio, in his recent book *Descartes' Error*, considers the necessity of emotional input for appropriate human thought and behavior, and for the direct perceptual sense of self. His terminology is somewhat different from that of James: he uses the word "emotion" for the unconscious reactive processes, and "feeling" for the conscious accompaniments. "If an emotion is a collection of changes in body state connected to particular mental images...*the essence of feeling an emotion is the experience of such changes in juxtaposition to the mental images that initiated the cycle.* In other words, a feeling depends on the juxtaposition of an image of the body proper to an image of something else, such as the visual image of a face or the auditory image of a melody." [3] From his studies of patients with various kinds of brain damage, he concludes that this inner sensation of bodily state is required for the experience of feelings, for appropriate behavior, and for the sense of self. Damage to part of the frontal lobes of the brain disconnects the inner sensations related to emotional states from the capacity for abstract and logical thought. These patients, though able to describe what one should or should not do under various theoretical circumstances, are notoriously inappropriate in their actual behavior, and concomitantly lacking in subjective feelings. Moreover, according to Damasio, background feeling, being "mostly about body states," [4] is the basis for individual identity. Self-consciousness is the result of the juxtaposition of internal and external sensation.

> I propose that subjectivity emerges...when the brain is producing not just images of an object, not just images of organism responses to the object, but a third kind of image, that of an organism in the act of perceiving and responding to an object. [5]

But Damasio leaves out one major fact: this kind of juxtaposition, resulting in self-awareness, is variable and discontinuous. Our waking state consists mostly of a kind of hypnotic sleepwalking, and the deliberate joining of body sensation and external sensation is the first step toward higher levels of consciousness. The importance of this fact was not lost on Ouspensky:

> I said that European and Western psychology in general had overlooked a fact of tremendous importance, namely, that *we do not remember ourselves,* that we live and act and reason in deep sleep, not metaphorically but in absolute reality. And also that, at the same time, we *can* remember ourselves if we make sufficient efforts, that we *can awaken.* [6]

This idea is found in many spiritual traditions, for example in the concept of mindfulness in Buddhism.

One could then provide a first answer to the question posed at the beginning: people fast, refrain from sex, and deliberately evoke fear and pain in themselves in order to produce bodily states, the awareness of which is one of the necessary components of a sense of self and of consciousness. These states ordinarily lead to actions which relieve them. If we had a well-developed capacity for self-awareness, whatever state the body was in would suffice as a component of consciousness. The Buddha, after trying all sorts of self-denials, gave all of that up when he became enlightened. But, for less developed beings, the intensification of body awareness created by deliberately provoking emotional-instinctive body activations and not allowing them to be relieved provides a necessary force for consciousness. One cannot so easily ignore these uncomfortable states and continue to dream.

The conglomerates (nuclei) of cells deep in the brain which organize our instinctive-emotional reactions also have a direct effect on the activities of the cerebral cortex. Anything that can be consciously perceived seems to require a representation in the cortex. The effect of these deeper unconscious structures on the cortex, aside from activating bodily responses which can then be perceived by sensation, consists of changing the *state* of the cortex in specific ways. At the grossest level, whether we are in deep sleep, in dreaming sleep, or awake, is regulated by cells that project diffusely to the cortex, giving rise not to specific sensations, but to changes in the connectivity patterns of cortical cells.

To clarify the concept of "state," one can consider the familiar states of matter: solid, liquid, and gaseous. A state, technically, is a possible arrangement of the components of a thing or a substance, as opposed to other possible arrangements. In each of the states of matter, the molecules of a substance remain the same, but their arrangement , and their relationship to each other, is different. The relationship of a substance to space and time changes radically depending on its state: solids move as a whole, liquids flow on surfaces, gases diffuse in whatever space is available. So does the relationship of a substance to other things. A hand can plunge through water or steam, but not through ice. Gases can intermix freely, liquids less so—alcohol and water intermix, but not oil and water—and solids do not intermix, though they can be permeated by gases and liquids. The state of a substance also changes its relationship to vibrations. Sound travels more quickly through water than through air. Water is relatively transparent to light, but both steam and ice are less so. And so on. So the state of a substance changes its relationship to space and time, to other things, and to vibrations.

Similarly, the nervous system can be in different states. The nervous system is composed of billions of massively interconnected nerve cells, or neurons. An average neuron receives thousands of contacts, synapses, from other neurons, and connects in turn to many others. Any particular neuron is entirely dispensable; only the activity of many, interacting with the environment and with each other, gives rise to perceptions, thoughts, memories, actions. A neuron sends more or fewer signals to other neurons depending on the summed input it itself receives from its synapses; synapses can be excitatory or inhibitory. But synapses are also modifiable; memories and habits reflect the activities of large ensembles of neurons whose connections have been strengthened, so that certain patterns of activity are easily triggered. And neurons can change their *state*, so that certain firing patterns are more or less likely to occur, and/or the neuron becomes more or less sensitive to some of its synaptic inputs. Some connections to neurons are not strictly excitatory or inhibitory, but modulatory; that is, they change the state of the neuron, making it more or less likely to respond to certain inputs or in a certain way. These state changes generally affect large ensembles of neurons together, or even the whole nervous system. Thus sleep, dreaming sleep, waking, and higher states of consciousness are mediated by changes in the connectivity patterns of the nervous system. This is reflected in the patterns of electrical vibratory activity of the nervous system, recordable as brain waves on the scalp. Like state changes in matter, these changes alter the relationship of our brains to the external world—to space and time, things and people, and external vibrations.

In my dreams I may fly through the air, but my body stays still; in fact it is prevented from responding to my mental images and emotional impulses by an active paralysis of the motor systems. In sleep, only a few external impressions reach my awareness; I am cut off from the world. And within each major change of state, there are many smaller ones. When I am hungry, the sight or smell of food takes on a different significance; food is perceived differently. If I am anxious and fearful, I am more likely to jump at a sudden sound, and regard other people warily.

The deep structures which regulate gross changes of state of consciousness, and those which mediate finer variations of state related to hunger, sexual desire, fear, anger, satiety, and all the other instinctive functions (in the Jamesian broad sense), are many and massively interconnected like all other parts of the nervous system. They could be considered to constitute an emotional-instinctive part of the nervous system, almost a separate brain, which, being subcortical, is not directly accessible to consciousness, although it regulates the states of consciousness of the cortex, and is in turn influenced by the activities of the cortex: by thoughts, memories, and perceptions. This part of the nervous system could be regarded as the physiologic substrate of the subconscious. It has its own perceptions, memories, and associations, as described in detail by LeDoux. Having been bitten by a dog as a young child, my fear of dogs may be entirely at odds with my conscious knowledge of them; I may not even remember the incident consciously.

Observe closely a discussion between two people. There can seem to be two conversations going on at once, one at an intellectual level, and the other emotional. Intellectually, there is the exchange of concepts, of logical points, of relevant information. Emotionally, the two react to each other: one may have a know-it-all tone, which irritates the other; or the other may be excitable and over-reactive, or on the other hand bored and distracted, which also provokes reaction. The two may actually agree in their thoughts, but fail to acknowledge it because of their reactions; or they may like each other and have compatible moods and feel they agree even though they are saying very different things. The emotional level of interchange may or may not be perceived. Sometimes the mind makes up reasons for disagreement when the emotions are actually the source. On the other hand, some people are very aware of this aspect and deliberately manipulate emotional reactions by flattery, or mood by plying with food, drink, and entertainment, to achieve agreement even when there is a fundamental difference of views.

The emotional brain is an organ of perception and communication. Fear is elicited, at least originally, by appropriate stimuli. Aggression is signaled by postures, facial expressions, gestures, kinds of vocalizations. So are a multitude of other emotional dispositions. Animals are constantly interacting in this way, and so are we, though with us these interactions are often obscured by layers of mental representations. The emotional brain is also best equipped to perceive subtle and rapidly changing facial expressions, postural tensions, and qualities of voice, which we may or may not be consciously aware of.[7]

Aside from provoking action, these perceptions make it possible to *know* how another person *feels*. For, just as I know how *I* feel at least partly as a result of bodily feedback, I can know how another feels by my reactions to his or her emotional communications. But this requires that these reactions be brought into the light, be illuminated by awareness, be coupled with thought. Otherwise they remain reactions. Thought and feeling perceptions need to face each other, rather than one being driven by the other. True imagination operates similarly: I imagine another's situation, and this evokes in me emotional responses, which when seen by consciousness allow me to feel how that person feels. This interaction, between thought, feeling, and awareness, makes conscience possible. Animals seem at least as capable as we at perceiving emotional communications, but lacking the kind of thought that enables us to imagine and foresee, they act without real conscience. Just as awareness of body state and external impressions together produces self-consciousness, a similar juxtaposition results in true consciousness of others, or conscience.

The retina, an outpost of the brain that has a multilayered neuronal structure like the cerebral cortex, takes the visual image projected onto it by the lenses of the eye, and transforms it by enhancing contrasts at the borders of visual objects and de-emphasizing gradual and mild variations of illumination within relatively uniform areas. These transformations are useful as a first step in breaking down the visual scene into objects. This function, contrast enhancement, is adjustable, by changing the connectivity patterns of retinal cells. In very dark conditions, contrast enhancement is no longer useful; rather, the retina must be tuned to signal any tiny spot of light that exists and not worry so much about shape. Thus the *state* of the retina changes in response to the needs of the visual system in different conditions of illumination. The changes in connectivity patterns of retinal cells are mediated by various neuromodulators within the retina, including dopamine.

Dopamine is also one of the neurotransmitters, or modulators, of the diffuse projecting systems that regulate our state of awareness and our emotional tone from the nuclei deep in the brain. There are a number of these diffuse projections. The neurotransmitters they use—dopamine, norepinephrine, serotonin, acetylcholine—are those whose functioning is altered by most of the mood-altering and psychedelic drugs. Dysfunction of these systems occurs in mental illnesses such as depression and schizophrenia. The same neurotransmitters also operate peripherally, in the body: acetylcholine and norepinephrine are the two major transmitters of the autonomic nervous system—of which the "solar," or celiac, plexus is a part—which regulate heart rate, blood vessel tone, sweating, intestinal activity, and many other functions related to the state of the body, the sensing of which seems to be a major part of knowing how we feel. In the brain, these neurotransmitters change the quality of perception by modulating the state of the cortex, its connectivity.

Aldous Huxley, after taking mescaline, looked at a vase of flowers on his desk "shining with their own inner light and all but quivering under the pressure of the significance with which they were charged. . . . A transience that was yet eternal life, a perpetual perishing that was at the same time pure Being, a bundle of minute, unique particulars in which, by some unspeakable and yet self-evident paradox, was to be seen the divine source of all existence."[8]

> If the doors of perception were cleansed every thing would appear to man as it is: infinite.[9]

Thus, changes in the state of the nervous system dramatically alter the quality of our perceptions, their *meaning*. Such changes can be induced by circumstances, or by practices which activate emotional energies and re-direct them so that they are not simply consumed by instinctive reactions.

Are these perceptual changes "real," a contact with a higher level of meaning, with the true significance of the universe, or are they delusional by-products of a purposeless evolutionary process? This question probably cannot be answered by arguments based on ordinary scientific observations, for perhaps true meaning is only perceivable when the nervous system has undergone this state change which arises from the joining of thought, feeling, and awareness. Many people find a meaning for life which transcends mere survival and egoistic concerns, whether it be faith in progress, or in one's country, or in mankind, or in love, or in the oneness of everything, or in God. For some, meaning is to be found in the grandeur of evolution itself. People hunger for meaning, more than for anything else, which is why they seek to transform their emotional energies from instincts for survival into a means of perceiving a higher purpose. But people differ tremendously in their interpretation of this sense of meaning, as suggested by the above list. Perhaps, just as the mind needs to be educated, so do the emotions, and even more the working together of the two, so that what is now only a momentary capacity for higher perception and consciousness followed by interpretations that are necessarily partial and faulty, becomes a consistently available faculty of perception. This is the science of the inner life which mystical traditions have sought to develop. One cannot make appropriate scientific observations without the necessary equipment.

> All our emotions are rudimentary organs of "something higher," e.g., fear may be the organ of future clairvoyance, anger of real force, etc.[10]

NOTES

1. Joseph LeDoux, *The Emotional Brain: The Mysterious Underpinnings of Emotional Life* (New York: Simon and Schuster, 1996).

2. William James, *The Principles of Psychology* (New York: Dover Publications, Inc., 1950), vol. II, pp. 449–450.

3. Antonio R. Damasio, *Descartes' Error: Emotion, Reason, and the Human Brain* (New York: G.P. Putnam's Sons, 1994), p. 145.

4. *Ibid.*, p. 155.

5. *Ibid.*, p. 242–243.

6. P.D. Ouspensky, *In Search of the Miraculous: Fragments of an Unknown Teaching,* (New York: Harcourt, Brace & World, Inc., 1949), p. 121.

7. Of course, detailed analysis of sensory impressions requires the cortex, and there is evidence that the non-dominant, or non-verbal, hemisphere is specialized for analyzing the emotional aspects of speech, for instance. How this hemisphere interacts with the subcortical emotional structures is of great interest. The whole question of the role and relationship of the two hemispheres of the brain, one more "rational" and verbal, and the

other more "holistic" and emotional, perhaps even more necessary for self-awareness, adds another important dimension to this discussion.

8. Aldous Huxley, *The Doors of Perception* (New York: Harper & Brothers, Publishers, 1954), pp. 17–18.

9. William Blake, *The Marriage of Heaven and Hell.*

10. G.I. Gurdjieff, *Views from the Real World* (New York: E.P. Dutton & Co., Inc., 1973), p. 194.

The Ego and the I

According to mystical perception, the universe is a great living organism. Like the human organism, which is analogous—created in the image of God— its life is self-sustaining by means of a vast and complex metabolic cycle, within which are many subcycles, wheels within wheels, levels within levels of interacting vibrations. Unlike the human organism, however, the universe doesn't depend on anything from outside, or perhaps God depends on the universe for His maintenance, the two being not the same and yet the same.

One half of the cosmic cycle is manifested as the outer world and is the subject of scientific scrutiny. It consists of many semi-independent entities on different scales: galaxies, stars, planets, animals and plants, nested within each other. It is a movement of creation and dispersal. According to science, all this began as a "big bang" from an infinitesimal point of infinite energy. Its overall direction is to run down, but in the process myriad transient self-sustaining entities are created, to live for a time. Some theorists believe that the apparent unidirectionality of time itself—broken glass does not spontaneously reassemble, and spilt milk doesn't gather itself back into the bottle, despite the basic laws of physics being time-reversible—derives from an extreme order at the beginning of time, which is gradually dispersing.

The other half of the cycle is not visible to science because it is manifest in the inner world of conscious beings. Consciousness reunites the dispersed elements, by the fact of their being perceived together. It counteracts time's arrow by remembering and anticipating. It is the way of return to the source.

> This is the truth: the sparks, though of one nature with the fire, leap from it; uncounted beings leap from the Everlasting, but these, my son, merge into It again.[1]

Thus God, or the universe, becomes tripartite:

> Bhartrprapanca...maintains that the selves and the physical universe are real, though not altogether different from Brahman. They are both identical with and different from Brahman, the three together constituting a unity in diversity.[2]

In the highest state of mystical perception a consciousness is united with everything, merges with universal consciousness; perceiver and perceived are one. Yet paradoxically this is not an annihilation of the self: the self and the greater Self of the universe are one, but the self retains its identity.

> He who understands, with the help of his personal self, merges himself into the impersonal Self.[3]

> How can even a mystic "die to self," and yet be conscious of God? Consciousness (irrespective of its object) implies continuation of self. A passing away of self cannot mean anything but sleep.[4]

The terms "ego" and "I" literally mean the same thing, but have come to connote quite different concepts, in a sense opposite, yet related. For human beings live in two worlds and have two kinds of selves. The first, lower self, is defined by the physical body in its relation to the external world. This body, being a temporary semi-independent entity which is part of the outward movement of creation, needs to maintain its integrity in the face of constant threat from destructive and dispersive forces—because it is part of the universal metabolic machine and was designed, via the process of evolutionary competition, to protect itself in order to fulfill its purpose. All the systems of the body serve to protect and maintain it. The nervous system, being the overall coordinator of the body's internal systems and external movements, provides the highest integrative level of protection by keeping the body focused on obtaining food, fighting or fleeing as necessary, and procreation so that more bodies will replace this one when it can no longer manage. The body's essential nature is to remain separate; it is part of the outward movement of differentiation and dispersal.

The ego, as Freud said, is "first and foremost a body ego."[5] Its origin is in instinctive self-preservation. Its excesses arise from the combination of the body's self-protective imperative and the uniquely human capacity for abstract thought and vivid imagination. These higher mental faculties make us exceptionally capable of defending ourselves, but overflow into the useless self-aggrandizing fantasies we are all familiar with.

The other, higher self is the self of the inner world, of consciousness. Its function is to reunite what has been separated, on many levels. On the practical level, consciousness unites materials to build structures; it creates order out of dispersed elements. On a higher level, somehow, mysteriously, it seems to reunite merely by perceiving. As with all metabolic cycles, this is not simply a return to the status quo; something has been gained, another drop of consciousness. In many religious traditions there is the idea that God created man so that He could contemplate Himself, thereby completing and maintaining Himself.

> He praises me (by manifesting my perfections and creating me in His form),
> And I praise Him (by manifesting His perfections and obeying Him).
> How can He be independent when I help and aid Him? (Because the Divine attributes derive the possibility of manifestation from their human correlates.)

For that cause God brought me into existence.
And I know Him and bring Him into existence (in my knowledge and contemplation of Him).[6]

A human being's perceptual capacities, if harmonized and honed, allow a merging of each individual consciousness with the consciousness of the universe. The essential nature of consciousness is to join. Yet it would seem to need first to be separate for this joining to be meaningful. A precursor chemical compound must be broken down before new compounds can be created. Thus the higher nature needs the lower, and individuation precedes transcendence.

The axis of consciousness is the I: "A thread that runs through me and every living being. One and multiple threads, one and multiple breathings. I am a cell in the organism of the universe. At a certain point the cell contains the universe."[7]

The same human capacities, which, properly harmonized and developed, thus can complete the cosmic cycle, when undeveloped and untempered by attention to reality join with the lower instinctive survival drive to create the hybrid monster of the ego.

In the physical world of space and things, the larger contains the smaller. It is a tenet of the laws of thermodynamics that the entropy, or disorder, of an isolated physical system must stay the same or increase; it can never diminish. So for a system, such as an organism, to become more ordered, it needs to borrow from the environment—overall, order cannot increase. The larger whole becomes more disordered, even though smaller elements within it can organize themselves. The universe, as a whole, must be running down. Yet, mystics perceive that the universe is alive and conscious, and at least self-sustaining. Where does its capacity to remain ordered come from? Order is synonymous with form, and form is created by consciousness. In the inner world of time and awareness, the smaller can contain the larger. Everything I perceive at this moment is contained in my inner world, and the present moment can contain all of time. There are hints of these mysteries in modern physics: in quantum theory the type of observation determines the form of what is observed, and in relativity theory the fourth dimension of time is of opposite sign to the three spatial dimensions[8]—but the mystery remains. Somehow the smaller—the individual consciousness—in joining with universal consciousness, imparts order to the larger world, preventing its decay over time.

> Smaller than the small, greater than the great, the self is set in the heart of every creature. The unstriving man beholds Him, freed from sorrow. Through tranquility of mind and of the senses (he sees) the greatness of the self.[9]

NOTES

1. *The Ten Principal Upanishads*, put into English by Shree Purohit Swami and W.B. Yeats (London: Faber and Faber Limited, 1952), p. 52.

2. *The Principal Upanishads*, edited with Introduction, Text, Translation and Notes by Radhakrishnan (London: George Allen & Unwin Ltd, 1953), p. 25.

3. *The Ten Principal Upanishads*, op. cit., p. 61.

4. R. Landau, *The Philosophy of Ibn 'Arabi* (London: George Allen & Unwin, Ltd., 1959), pp. 51–52.

5. Sigmund Freud, quoted in Oliver Sacks, *The Man who Mistook his Wife for a Hat* (New York: HarperCollins Publishers, 1990), p. 52.

6. R. Landau, op. cit., p. 74.

7. Michel de Salzmann, unpublished remarks.

8. In space, the distance between two points, by the Pythagorean theorem, is $\sqrt{x^2+y^2+z^2}$. In four-dimensional space-time, the interval between two points is $\sqrt{t^2-x^2-y^2-z^2}$.

9. *The Principal Upanishads*, op. cit., p. 617.

THE HOME OF THE SELF

A home is much more than a physical dwelling. The expression "he is never at home" can refer to someone who is always out of his house, or to someone who is constantly distracted. "Out to lunch" has the same dual meaning. And one can say "she is (or is not) at home in her own skin." In French, there is no noun for home, but "chez moi" means "where I dwell," both physically and psychologically.

Home is the place where the world and I meet. It is not isolated from the outside, nor impermeable to it, but neither is it indiscriminately open. There are walls and a roof, but also doors and windows, and nowadays other portals: plumbing, electric and telephone lines. I invite people into my home, but not just anybody. I choose who and what to let in, and also what goes out. My guests leave with selected impressions; I do not show them everything.

Similarly the contents of my home are selected. I bring in furniture, equipment, supplies, and decorations, and arrange and modify them to suit my needs and tastes. A hotel room is not home, though it may have the basic necessities, because its contents have not undergone this transforming process.

A well-functioning home is like an organism. An organism is a highly organized, relatively independent entity, in dynamic relationship with its environment. It maintains its form, its life, by selectively taking in the materials it needs and transforming them into its constituents, and excreting things it does not need, or exchanging them for those it does. These materials are of different densities: there is ordinary food of various kinds, water, and air, and most important, sense impressions, which allow it to find food and avoid danger. An organism is isolated from the outside by its skin, but it has doors and windows through which it selectively regulates its input and output.

A cell has a similar organization. Relatively independent, its status as an entity is defined by its membrane. This membrane is selectively and actively permeable, allowing in necessary biochemical constituents to maintain the life of the cell, and specific amounts of water and ions to keep its shape and provide the electrical potentials that power it.

The home, the organism, and the cell are all hierarchical organizations. The nucleus provides the instructions for the activities of the cell, but these instructions are flexible, responding to the needs of the cell and to its environmental interactions. In the organism, the nervous system is the primary organizer, although the immune and other systems also play a role. In the home, the people determine the contents of the home and its relationships with the outside world.

Each of these entities also contains smaller entities. The home contains the people who live there, and also makes use of hired help: plumbers, cleaning people, etc. The organism is composed of cells, which serve the overall organism while maintaining themselves. In addition, intestinal and other bacteria serve useful functions, although they are not considered part of the organism. Even the cell has "hired help": the mitochondria, subcellular organelles that are the main energy factories of the cell, are thought to have once been independent organisms.

Each relatively independent entity can be regarded as a cosmos. We are familiar with the idea of the macrocosmos, the universe, and the microcosmos, man. In this traditional view, human beings are considered to be miniature replicas of the whole. But according to some, there is a hierarchy of cosmoses,[1] each similar to the whole, each an organism in its own right: the universe, the galaxy, the solar system, the planet, the animal, the cell. In this view, the universe consists of a nested series of interacting, interdependent organisms, and this is its basic organizational pattern. Each cosmos maintains its identity, its life, for a time, by virtue of its self-generated selective relationship to the outside world, and also serves the purposes of the larger entities it is a part of.

While we take for granted that such hierarchical arrangements of nested entities are the norm in the human realm—countries, states, municipalities, organizations, homes—we do not generally perceive the whole universe to be organized in this way.

G.I. Gurdjieff said: "Knowledge begins with the teaching of the cosmoses,"[2] emphasizing the importance of this organizational principle.

In this view also, human beings have a special role, by virtue of our capacity for consciousness, but we do not necessarily fulfill it. While the body is a well-functioning cosmos, at least for a time, the psyche generally is not. Our inner lives are buffeted here and there by random associations and reactions to external stimuli. We start out with a particular purpose, and wind up creating the opposite effect of what we intended. A perceived criticism changes our mood for the day. We tend to live in an inner chaos of daydreams, partially digested notions, and half formulated plans. But the possibility exists for the development of attention, awareness, and intentionality, for the growth of presence. People vary tremendously in this respect. We have all met people who are so distracted they cannot accomplish any task, and others who are to various degrees simultaneously within themselves—at home—while interacting with and manifesting purposefully in the world. In the end a finer body of attention, of presence, can exist within the physical body. It has all the characteristics of another cosmos: it is relatively independent, self-contained, and selectively permeable to the outside. While the nature of the materiality of this entity is unclear, for some this is the soul, which can exist independently of the physical body if it is sufficiently formed. This body of attention and awareness may play a different role from the physical body in the larger cosmoses of which it is a part. The physical body is part of the ecology of the earth, transforming biochemical substances; the soul is part of the ecology of consciousness, transforming perceptions.

Swing low, sweet chariot
Coming for to carry me home.

The real meaning of coming home to God may be the establishment of a home in oneself, which permits one to share in, and contribute to, the universal consciousness.

He through Whom we see, taste, smell, feel, hear, enjoy, know everything, He is that Self.

Knowing That by Which one perceives both dream states and waking states, the great, omnipresent Self, the wise man goes beyond sorrow.

Knowing that the individual Self, eater of the fruit of action, is the universal Self, maker of past and future, he knows he has nothing to fear.

Born in the beginning from meditation, born from the waters, having entered the secret place of the heart, He looks forth through beings. That is Self.

That boundless power, source of every power, manifesting itself as life, entering every heart, born with the beings, that is Self.[3]

NOTES

1. P.D. Ouspensky, *In Search of the Miraculous: Fragments of an Unknown Teaching* (New York: Harcourt, Brace & World, 1949).

2. *Ibid.*, p. 205.

3. *Katha Upanishad*, edited by the author based on multiple translations.

4

WORLDS WITHIN WORLDS

THE TEACHING
OF THE COSMOSES

In recent years, there has been an increasing interest in reconciling, or at least understanding the relationship between, science and spirituality. Neuroscientists are tackling the question of the neural correlate of consciousness, after avoiding the subject for a long time. Philosophers are seriously studying the sciences. Physicists find themselves pondering the relationship between their theories and age-old spiritual questions. Understanding the nature of the world and our place in it has always been the goal of both the study of the external world and the inner search for meaning, but in modern times these two approaches became artificially separated, almost as if to give the powerful methodology of science a chance to develop. Now, however, it seems time for attempts at reunification.

This has by no means been achieved. So, if a spiritually minded thinker, with an interest in and respect for science (there are still many on both sides who simply dismiss the other point of view), is asked: "What is the difficulty?", a few answers repeatedly appear.

One is that science is an outer pursuit, dependent on objectively verifiable experiment on the material world, while spirituality is an inner pursuit, consciousness being inherently subjective. This was summed up by Schrödinger in his "principle of objectivation": "A physical scientist does not introduce sensation or perception into his theories. Having removed the mind from nature, he cannot expect to find it there." [1] There are many versions of this answer: science is not concerned with meaning, or values, or even the question "why?" These variations are all related, and reflect the difficulty of finding the intersection of spirit and matter, of the inner and outer worlds.

Another recurring answer to the question is that science regards everything as being on the same level, made of the same stuff and subject to the same laws, whereas spirituality recognizes a hierarchy of levels, from the fine to the coarse, from spirit to matter, from God to humankind. Most often, two levels are specified: the macrocosmos and the microcosmos, God and humanity, or God and Nature, thought to be intimately related, or even self-similar, yet clearly in a hierarchical relationship.

"God created Man in His image."
"As above, so below."

More elaborate expositions recognize more than two levels, or cosmoses, often seven. This view has been put into relatively modern form by G.I.

Gurdjieff, and is the basis of his worldview:

"Knowledge begins with the teaching of the cosmoses." [2]

But surely science takes into account huge differences in, or levels of, the size of entities—from quarks to atoms to organisms, planets, stars and galaxies—and corresponding differences in the speed of interactions. Is there a difference between these levels, recognized by science, and the teaching of the cosmoses? Is it merely a difference in terminology? In scientific parlance, "cosmos" is singular, synonymous with the universe, studied by "cosmology". What, in Gurdjieff's view, and that of ancient teachings, is a cosmos?

Gurdjieff proposed seven cosmoses, given somewhat different definitions and names in the two books in which this idea is treated in some detail: *In Search of the Miraculous*, a reportage by P.D. Ouspensky of Gurdjieff's teaching, which Gurdjieff himself said accurately reflected his teaching from 1914 to 1917, and *All and Everything, First Series, Beelzebub's Tales to his Grandson*, written by Gurdjieff in the 1920s and 30s. There are some slight discrepancies, such as whether the planets of a solar system constitute a separate cosmos from the solar system itself, but one can with reasonable confidence identify the following as cosmoses:

God
The universe
The galaxy
The star, sun, or solar system
The earth, or planet
The multicelled organism, and human organisms in particular
The cell, or "microbe" as it is put in *In Search of the Miraculous*

Ouspensky went on to consider the molecule and the electron to be cosmoses, and classified small and large cells as separate cosmoses, but these ideas did not originate with Gurdjieff. Ouspensky went further and produced elaborate tables relating cosmoses to each other in an incorrectly precise way. *All and Everything*, on the other had, is written in a deliberately obscure manner. Despite these difficulties there is, I think, something fundamental in the idea of cosmoses.

I do not wish to quote the expositions in these two books in their entirety, although they are well worth reading, but certain clues stand out regarding the nature of a cosmos:

A cosmos is a "relative independent concentration." [3]

"Each cosmos is a living being which lives, breathes, thinks, feels, is born and dies."

[Cosmoses are] "included one within another and created in the image and the likeness of the greatest of them, including in itself all the others. 'As above, so below,' in an expression which refers to cosmoses."

"The idea of [only] two analogous cosmoses...is so incomplete that it can give no idea whatever of the analogy between man and the world."

"All cosmoses result from the action of the same forces and the same laws...But they manifest themselves in a different, or at least, in not quite the same way on different planes of the universe, that is, on different levels...Owing to the law of octaves there is no complete analogy between them, just as there is no complete analogy between the different notes of the octave. It is only *three* cosmoses, taken together, that are similar and analogous to any other three."

"The idea of the possibility of broadening man's consciousness and increasing his capacities for knowledge stands in direct relation to the teaching on cosmoses. In his ordinary state man is conscious of himself in *one cosmos*... The broadening of his consciousness and the intensifying of his psychic functions lead him into the sphere of activity and life of *two other* cosmoses simultaneously, the one above and the one below, that is, one larger and one smaller."

"The manifestation of the laws of one cosmos in another cosmos constitutes what we call a *miracle*...A miracle is not a breaking of laws, nor is it a phenomenon outside laws. It is a phenomenon that takes place according to the laws of another cosmos."

"It is very useful to examine the life and phenomena of one cosmos as though looking at them from another cosmos...All the phenomena of the life of a given cosmos, examined from another cosmos, assume a completely different aspect and have a completely different meaning. Many new phenomena appear and many other phenomena disappear. This in general completely changes the picture of the world and of things."

"One cosmos is related to another as *zero to infinity*...In order to understand the meaning of the division into cosmoses and the relation of cosmoses to each other, it is necessary to understand what the relation of zero to infinity means. If we understand what this means, the principle of the division of the universe into cosmoses, the necessity of such a division, and the impossibility of drawing for ourselves a more or less lucid picture of the world without this division will immediately become clear to us."

"The idea of cosmoses helps us to understand our place in the world; and it solves many problems, as for instance, those connected with space, with time, and so on. And above all this idea serves to establish exactly the *principle of relativity*."[4]

Following his exposition, as reported in *In Search of the Miraculous*, Gurdjieff asked Ouspensky to explain his concept of dimensions. This began with the standard mathematical definition of dimensions (point, line, plane, etc., each related to the one above as zero to infinity, but included three, rather than one, time dimensions.[5] At the end of this, Gurdjieff said:

"There is a great deal of material in what you have just said, but this material must be elaborated. If you can find out how to elaborate the material that you have now, you will understand a great deal that has not occurred to you till now. For example, take note that *time* is different in different cosmoses. And it can be calculated exactly, that is, it is possible to establish exactly how time in one cosmos is related to the time of another cosmos.

"I will add only one thing more:

"Time is breath — try to understand this."

Ouspensky reported that later Gurdjieff said that the sleep and waking of living beings and plants, a day and a night, constituted the "breath of organic life".[6]

"...the emanations and radiations issuing from all these cosmoses of different scales, by means of which the process of the great cosmic Trogoautoegocrat [the reciprocal feeding or exchange of substances by which the universe maintains itself] proceeds."[7]

By the early nineteen-teens, when Gurdjieff began teaching in Russia, Einstein had published his special theory of relativity and was working on the general theory, and the quantum picture of the atomic world was emerging. It is unclear to what extent Gurdjieff was acquainted with these developments, but his ideas were certainly not based on them, so some of the relationships between Gurdjieff's ideas and the revolutions in modern science are all the more striking. Relativity theory overturned the Newtonian picture of an unchanging background framework of the world, consisting of three perpendicular spatial dimensions, and an independent single dimension of time, flowing steadily from past to future, in which all events took place without affecting the structure of the framework. Special relativity dictated that space and time were interrelated, and that both were affected by relative motion, so that every moving object has its own spatiotemporal coordinates, length in the direction of motion being shortened in proportion to the speed of motion, and time being expanded, as viewed by an outside observer. These requirements followed simply from the two basic postulates of special relativity, that the laws of nature are the same in all non-accelerating reference frames (a ball dropped in a smoothly moving train falls straight down, and all other motions within the train appear normal to the riders), and that the speed of light and of other electromagnetic waves is constant, regardless of the motion of the perceiver. The results are astounding, and amply confirmed:

nothing can travel faster than light, because if an object were to travel at lightspeed, its thickness would be zero and its mass infinite. Energy and mass are convertible one into the other, because the energy provided to make an object move faster is reflected in an increase in its mass proportional to its velocity. And for light neither space nor time as we perceive them exist. This certainly suggests that light lives in a different dimension. General relativity further undermined the Newtonian framework by making the shape of space dependent on the masses within it, the distortion of space being the cause of gravitational effects.

In quantum theory, further peculiarities regarding space and time became evident. By Heisenberg's uncertainty principle, the exact position and exact momentum of a particle cannot be simultaneously observed, nor can the precise energy of an event and the exact time when it occurred. It is as if we were confined to a particular plane of observation at any given time, and could not make observations from a perpendicular viewpoint. Pairs of parameters, position and velocity, energy and time, are then in some sense perpendicular to each other. Similarly, the dual wavelike and particle-like nature of the fundamental building blocks of matter—electrons, photons, atoms—cannot be observed simultaneously, the nature of the manifestation depending on the choice of observation, raising the question of the role of consciousness in the world.

The theories of relativity and the quantum theory enabled tremendous advances in the ability of physics to explain phenomena, but in the process the concepts of space and time, matter and energy, and their relationships, became completely different from those we develop naturally as a result of our ordinary interactions with the world. A further difficulty is that Einstein's theory of gravity and the quantum theory of the other fundamental forces (electromagnetism and the weak and strong nuclear forces) remain irreconcilable. Attempts to reconcile them are very active today, and the resulting tentative theoretical frameworks wreak further havoc with our ideas about space, time and matter: ten and eleven dimensional "string theories" and the like. The jury is still out.

But it remains clear that neither space nor time can be regarded as an independent background for events, nor can the observer and the observation be completely separated. Lee Smolin, in his book on current thinking about attempts to reconcile quantum theory and relativity theory, *Three Roads to Quantum Gravity*, points out that the irreconcilability of the two theories is intimately related to these basic questions:

"In this book we shall be concerned with two very basic ways in which theories may differ. The first is in the answer they give to the question of what space and time are...Another way...is in how observers are believed to be related to the system they observe.

"The problem is that while quantum theory changed radically the assumptions about the relationship between the observer and the observed, it accepted without alteration Newton's old answer to the question of what space and time are. Just the opposite happened with Einstein's general theory of relativity, in which the concept of space and time was radically changed, while Newton's view of the relationship between observer and observed was retained. Each theory seems to be at least partly true, yet each retains assumptions from the old physics that the other contradicts."[8]

Gurdjieff's view of time is not unlike that of Smolin and some other modern physicists. In *Beelzebub's Tales*, in the chapter "The Relative Understanding of Time", he says:

"Time in itself does not exist; there is only the totality of the results flowing from all the cosmic phenomena present in a given place...It does not issue from anything, but blends always with everything and becomes self-sufficiently independent; therefore, in the whole of the Universe, it alone can be called and extolled as the 'Ideally-Unique-Subjective-Phenomenon.'"[9]

Smolin's view may appear different on the surface, but on closer examination is quite similar: he argues that neither space, nor time, nor things, are fundamental, but that the basic building blocks of the universe are processes, which can be regarded as information exchanges and transformations ("processing").

"Such a universe has time built into it from the beginning. Time and change are not optional, for the universe is a story and it is composed of processes. In such a world, time and causality are synonymous. There is no meaning to the past of an event except the set of events that caused it. And there is no meaning to the future of an event except the set of events it will influence...A causal universe is not a series of stills following on, one after the other. There is time, but there is not really any notion of a moment of time. There are only processes that follow one another by causal necessity."[10]

"So there are not really two categories of things in the world: objects and processes. There are only relatively fast processes and relatively slow processes."[11]

Or, more simply: "Time is nothing but a measure of change — it has no other meaning. Neither space nor time has any existence outside the system of evolving relationships that comprises the universe."[12]

Smolin goes on to discuss the scale at which time and space are postulated to become discrete and discontinuous, the Planck scale, of the order of 10^{-33} centimeters, 10^{-43} seconds, and at an astronomically high temperature. "Our world is incredibly big, slow, and cold compared with the fundamental world."[13]

Gurdjieff: "It is necessary to notice that in the Great Universe all phenomena in general without exception wherever they occur and manifest, are simply law-conformable 'fractions' of some whole phenomenon which has its prime arising on the 'Most Holy Sun Absolute'". [14]

These are two clear statements of scale, in very different terms. But Gurdjieff takes a further step:

"The chief particularity of the process of the flow of Time in the presence of cosmic arisings of various scales consists in this, that all of them perceive it in the same way and in the same sequence." [15]

He then goes on to describe the experiences of "microcosmoses", or single celled organisms:

"These infinitesimal beings also, like the beings of cosmoses of other 'scales', have their experiences of a definite duration for all their perceptions and manifestations; and, also, like them, they sense the flow of Time by the comparison of the duration of the phenomena around them.

"Exactly like the beings of other cosmoses, they are born, they grow up, they unite and separate for what are called 'sex-results' and they also fall sick and suffer, and ultimately like everything in which Objective Reason has not become fixed, they are destroyed forever.

"For the entire process of the existence of these infinitesimal beings of this smallest world, Time of a definite proportionate duration also ensues from all the surrounding phenomena which are manifested in the given 'cosmic scale'.

"But actually from the point of view of the sensation of the duration of Time by your favorites of the planet Earth, the whole length of the existence of the 'beings-Microcosmoses' lasts only a few of their 'minutes' and sometimes even only a few of their 'seconds'." [16]

This is the one of the main differences between the concept of simply "scale", and the idea of "cosmos". A cosmos has a self-contained reality in addition to its relational reality to other scales or cosmoses. It perceives time "in the same way and in the same sequence". It is a world of its own. This is the only conceptual framework in which it is possible to conceive that God made man in his own image.

To understand the nature of a cosmos, the cell is a good place to start. A cell can survive on its own, given the proper environment. It is separated from the world around it by a membrane, which does not keep out all outside forces, but is actively semi-permeable. The cell selects which substances it will allow in or out; it takes in nourishment of various kinds, and secretes its wastes, as well as substances needed for other cells or for the environment (including the larger organism it may be part of), according to its particular role. The nourishment it needs is of various kinds. It needs organic molecules to provide its energy, and needs to allow certain ions to enter more readily than others in order to maintain an electrical potential across its membrane. This electrical potential is essential for its various functions and for its semi-independent existence as an entity, and, along with the electrochemical energy obtained from its food, could be called its life energy. A dead cell has lost its membrane potential and its relative autonomy. There is a hierarchy within a cell: its DNA provides instructions for the assembly of its constituents, and subsets of instructions are selected for implementation on a continually changing basis, according to the cell's needs, changes in the environment, or requests from the environment. These interactions could be regarded as its brain, and the proteins that make up its structure and provide the molecular tools for its functions its body.

An animal is also a cosmos, composed of cosmoses of the next level, cells. It is born, lives, and dies. It is relatively independent and maintains a selective separation from its environment, allowing certain substances in (food, air), and others out (excretions, sweat, deoxygenated air). But it has another level of intake and output: it takes in impressions from the environment and excretes behavior, according to its role in nature.

Human beings, endowed with a brain that can comprehend abstract ideas, also take in and excrete ideas, feeding each other. While the intake of food and air is carefully regulated automatically by the body, the intake of impressions of the world, their comprehension (metabolism) and the resulting output of ideas and behavior requires a non-automatic active and honed attention for there to be an appropriate level of functioning for a human being. As we mostly are, we squander our capacities in egotistic pursuits and conflicts. According to Gurdjieff, we have a potentially greater role than other animals in the universal exchange of substances. Our consciousness feeds other cosmoses, just as the membrane potentials of nerve cells feed our brains, by providing the signals needed for sensation, perception, movement and thought. To fulfill this greater role, human beings need a semi-permeable membrane on another level, maintained by an active, conscious attention, completely analogous to skin and the mucous membranes of the lung and intestine. As we are, we are mostly cosmoses on one level only; potentially, we are cosmoses on another level. According to Gurdjieff, this higher cosmos, if sufficiently developed, can take on an independent existence, as the soul, and survive the death of the body, having created a new structure, made of finer materiality. Souls fulfill the same function in the greatest cosmos as the cells of our brains fulfill in us.

"This localization, which is concentrated in their head, they call the 'head-brain.' The separate, what are called 'Okaniaki' or 'protoplasts' of this localization in their head, or as the terrestrial learned call them, the 'cells-of-the-head-brain,' actualize for the whole presence of each of them exactly such a purpose as is fulfilled at the present time by the 'higher-perfected-bodies' of three-brained beings from the whole of our Great Universe, who have already united themselves with the Most Most Holy Sun Absolute or Protocosmos."[17]

The earth is also a cosmos. It is composed of many things, but on its surface at least, it is an aggregation of lower cosmoses, microcosmoses and multicelled organisms. Hundreds of billions of cells make up our bodies, and hundreds of billions of organisms make up the surface of the earth. The earth also takes in food, from the sun, but selectively. The atmosphere only allows certain frequencies of electromagnetic radiation to reach the surface. The sun also interacts with the earth through the earth's electromagnetic body, the magnetosphere, which also is selective in relation to the solar wind, although its function is not well understood and its complexity only beginning to be appreciated. So the earth also has a semi-permeable membrane. As with a cell membrane, this creates a separation of charges: there is a progressive voltage difference between the surface of the earth and the atmosphere of 100 volts per meter, replenished periodically by lightning.[18] The earth's electromagnetic body could well interact with ours, resulting in an exchange of subtle energies, and the earth's electromagnetic body also interacts with the sun's, and with the moon, which passes within the tail of the earth's magnetosphere every time it is full. According to Gurdjieff:

"...all the results obtained from transformations in different cosmoses localize themselves together according to what is called 'qualitativeness of vibrations,' and these localizations penetrate everywhere throughout the Universe and take a corresponding part in planetary as well as in surplanetary formations, and generally have as the temporary place of their free concentration the what are called atmospheres, with which all the planets of our Megalocosmos are surrounded and through which connection is established for the common-cosmic Iraniranumange [exchange of substances]."[19]

One of Gudjieff's main tenets is that organic life on earth serves in part to feed the moon, giving it life over a long period of time. This is of course well beyond the ideas of science at the present time, but if it is true, a mechanism exists.

What then is a cosmos? A possible definition: a cosmos is a relatively self-contained and relatively independent entity, with a life of its own, which contains within itself a representation of the basic laws that govern processes, and which, in order to maintain its relative independence, is separated from everything else by a semi-permeable, or selectively permeable, membrane, through which it obtains what it needs, and interacts with other cosmoses, both on the same and on other levels.

Gurdjieff stated that cosmoses are related to each other as zero to infinity, and emphasized that it is necessary to understand what this means, and that the idea of cosmoses could clarify many problems, including those related to understanding space and time. He then let Ouspensky discuss the mathematics of dimensions. Mathematically, one dimension is related to another as zero to infinity. An infinite number of points (a mathematical point having no size whatsoever, only a location) fit into a line, and infinite number of mathematical lines (which have length but no thickness) fit into a plane, etc. These abstract and ideal concepts have served well in developing the mathematical tools that are used to understand physical phenomena, and the correspondence between mathematical ideas and the laws that govern the universe is uncanny and mysterious. However, there is no actual physical point which has no size at all, and no physical line which has no thickness. This in fact gives rise to one of the difficulties which exists in modern physics, and is part of the problem in reconciling quantum mechanics and relativity theory. In quantum theory, fundamental particles are treated as mathematical points, and as a result if one measures the electric field arbitrarily close to an electron, for instance, the field becomes infinite. To get around this problem, physicists have had to resort to a mathematical trick, called renormalization, which works well—quantum theory being one of the most precise theories known for calculating the results of interactions—but has left many physicists uneasy because of the arbitrariness of the procedure. Einstein's theory of gravity cannot be renormalized, so it cannot be reconciled with quantum theory.[20] That is one of the reasons why many of the current attempts to reconcile the two theories postulate that space and time are not infinitely divisible: on the smallest scales, space and time must be discrete, so that a dimensionless point cannot exist.

In fact, the problem of regarding time and space as infinitely divisible caused consternation among the ancient Greeks, as exemplified in Zeno's paradoxes. One of them states that motion is impossible, because for an object to move from point A to point B it must first move to point C, between A and B, and to move to point C, it must first move to point D, between A and C, and so on ad infinitum, since there are an infinite number of points between A and B. So things cannot move at all, since an infinite time would be required to move through an infinite number of points. These difficulties were resolved mathematically by the calculus, which made it possible for Newton to calculate motion, and ushered in the modern scientific era. But the calculus also relies on a trick, which depends on squared infinitesimals being ignored altogether, so that "regular", non-squared, infinitesimals can be compared with each other, giving finite results.

Gurdjieff's statement that one cosmos is related to another as zero to infinity makes the concept of a dimension quite different from the mathematical one. A cell is much much smaller than a human being, but it is not a mathematical point. An animal is much smaller than the earth, but is also not a mathematical point. It is a curious fact, made much of by Ouspensky, that for our perception cells (some of the largest being exceptions) do effectively have zero size, being just beyond the limit of visibility. Similarly, the earth is effectively almost flat for our perception; perceptually we usually live on a plane. The sun's motion in the sky, caused by the earth's rotation, is just beyond the limit of our ability to perceive motion, except when it is on the horizon, when its motion can just barely be perceived. So, for our perception, the lower cosmos and the higher one are of a different dimensionality compared to ours.

Related to change in dimensionality is change in the nature of phenomena, when viewed from different levels, as noted by Gurdjieff. A single blood, lung, or kidney cell is of no importance to any of us, although it contains a whole world in itself. But the millions of cells which together make up the blood, lung and kidney perform vital transformative functions in our bodies. Most likely any human being is not important for the earth, but humanity as a whole may perform a vital function. With a scratch we heedlessly kill many skin cells; earthquakes and volcanoes can do the same to us. With a change of scale there is also a change from the discrete to the continuous, and from the particular to the statistical. Individual molecules in a gas are bouncing off the walls of the container and each other in specific discrete ways , but the gas as a whole has a smoothly varying pressure and temperature. I do not know when I will die, but the insurance companies, looking at large populations, can determine quite well how many people of a given age will die in a given time.

A rapidly rotating propeller acts as a solid object. The objects around us, viewed at the atomic level, consist of very rapid vibrations. In a musical tune, the notes are discretely placed within the rhythmic structure, but on another level each note is itself a rhythmic vibration.

So the dimensions of Gurdjieff's system of cosmoses are not quite the same as abstract mathematical dimensions. Looked at from our scale, the cell is *almost* a point, the earth is *almost* flat, and the speed of light is *almost* infinite. We could postulate that the real dimensions of the world are *almost* perpendicular. Einstein's spacetime is not Euclidian: its shape is distorted locally by mass. The constancy of the speed of light, regardless of the velocity of the reference frame from which it is viewed, is analogous to the constancy of the distance from the surface of a sphere to the center, regardless of one's movement on the surface: light comes from another dimension.[21] If the real dimensions that separate cosmoses were truly perpendicular, and the relationship of one cosmos to another were mathematically zero to infinity, there could be no interaction between cosmoses.

Another major difficulty in reconciling Einstein's relativity theory and quantum theory, as pointed out by Smolin, is in the different perspective of each theory with regard to the relationship between the observer and the observed. This reflects the quandary pointed out at the beginning of this essay: the difficulty of finding the intersection of the outer world and the inner life. For the inner life cannot be found in space. No amount of probing the brain with microscopes or electrodes or MRI scanners reveals it, although physical correlates can be found. For some, this leads to its dismissal as an epiphenomenon, whatever that means; it is somehow not real. But since the inner world is the only world we actually know, others have dismissed the outer world in a similar way, as an illusion. If consciousness is an important ingredient of reality, the inner world cannot be dismissed, nor, according to our common sense, can the outer. They must both be equally real.

If the inner world is not present in space, where is it? According to some, the inner world exists only in the dimensions of time—Gurdjieff's "Ideally-Unique-Subjective-Phenomenon"—which dimensions must be more than singular to support the vastness of the inner world. Ouspensky postulated three dimensions of time, discussed elsewhere.[22] What I would point out here is that what is "inner" on one level, in one cosmos, may become "outer" on another. The totality of people's perceptions, thoughts, and feelings may serve as food, consisting of vibrations, synonymous with substance, for the earth and the solar system.

According to Gurdjieff, human beings as animals are examples of a cosmos on a level between planet and cell, but, as touched upon above, because of the capacity given to us by our having three brains—intellectual, emotional, and moving-instinctive—unlike other mammals which have only two—we can develop in ourselves cosmoses on other levels. The addition of the third brain, the intellectual brain, capable of abstract perception, makes possible, when it functions in harmonious relationship with the other two, a larger correspondence with reality, a resonance with a higher level of consciousness. For our ordinary animal life, we depend on substances from the earth alone, but by absorbing substances from the level of the solar system, another body can be formed within us, and within that second body, a third can be formed from the absorption of substances directly from the highest level.[23] These concepts are certainly foreign to science, and well beyond the direct knowledge of most of us, so that one hesitates to even mention them, but they are at the core of Gurdjieff's teaching. However, we certainly experience changes of consciousness, from sleep, to waking—itself a semi-dreamlike state—, to moments of a clear and vivid impression of oneself in the world, in the present. In these moments, the world acquires a vividness unknown before, and the sense of time changes dramatically.[24] One aspect of this change is an expansion of consciousness to include a higher and a lower cosmos, as described by Gurdjieff.

This change is a change of *state*, which is a change of level, and of dimensionality. In physical science, a change of state, as from ice to water to steam, is a change in the arrangement and relationship of elements, in this example the molecules of H_2O. It involves a change in the relationship of these elements to the dimensions of spacetime. A block of ice can only move in a line, at any given time (even though the line can be curved in all three spatial dimensions). Water moves in a plane; steam in three dimensions at once. A change of state also changes the relationship of a substance to other influences. Light does not penetrate far into ice, but much further into water. A static charge only feels electrical influences, a moving one becomes sensitive to magnetic fields (a static charge will feel magnetic influences from a moving magnetic field; it is the relative motion which is important).

These considerations are not in the language of hard science. Whether they can be put into a scientific mathematical framework remains to be seen. They seem to bear rather directly on some of the unsolved questions of physics. But there is not an obvious correspondence between Ouspensky's three space and three time dimensions and the latest superstring theories of nine or ten spatial and one time dimension. (However, at least one prominent physicist, Stephen Hawking, has postulated a second time dimension.) For many who have worked with Gurdjieff's methods, his ideas make eminent sense, based on their experiences, although the full range of Gurdjieff's worldview is not so accessible. Perhaps over time Gurdjieff's ideas will help to bring about an exact science which includes the inner and outer worlds, time and space and things, consciousness, energy and matter.

NOTES

1. Gerald M. Edelman and Giulio Tononi, *A Universe of Consciousness: How Matter Becomes Imagination* (New York, Basic Books, 2000), p. 217.

2. P.D. Ouspensky, *In Search of the Miraculous: Fragments of an Unknown Teaching* (New York, Harcourt, Brace &World, Inc., 1949), p. 205.

3. G.I. Gurdjieff, *All and Everything, First Series, Beelzebub's Tales to his Grandson* (New York, E.P. Dutton & Co. Inc., 1950), p. 759.

4. Ouspensky, p. 205–208.

5. Ouspensky, p. 208–213.

6. Ouspensky, p. 213.

7. Gurdjieff, p. 760.

8. L. Smolin, *Three Roads to Quantum Gravity* (New York, Basic Books, 2001), p. 4.

9. Gurdjieff, p. 123–124.

10. Smolin, p. 54–55.

11. Smolin, p. 52

12. Smolin, p. 24.

13. Smolin, p. 63

14. Gurdjieff, p. 123.

15. Gurdjieff, p. 125.

16. Gurdjieff, p. 125–127

17. Gurdjieff, p. 777–778.

18. Wikipedia, article on "Atmospheric Electricity."

19. Gurdjieff, p. 763.

20. Smolin, p. 156.

21. See "The Light of the Beholder" in this volume.

22. See "Does Man Have Three Brains?" in this volume.

23. Gurdjieff, p. 569.

24. See "Shadows of the Real World" in this volume.

HOLY EARTH
Is the Earth a Conscious Being?

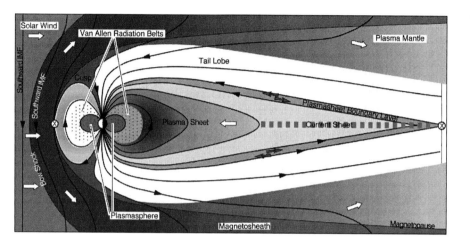

Is the earth conscious? Surely one of the attributes of holiness is awareness. A holy man cannot be a robot; on the contrary, his awareness is assumed to be greater, or more comprehensive, or on a different level, than that of an ordinary person. And it is our sense of the holiness of all that exists that necessitates the existence of a conscious God or gods, regardless of the forms in which we conceive of them.

To consider this question with some rigor we need to delineate the elements that are required for consciousness to exist, and see whether the earth as a whole meets those requirements. This is a difficult, perhaps even arrogant, inquiry. It implies that consciousness can be studied scientifically like any other natural phenomenon, and this is a subject of considerable debate. Furthermore, although consciousness *is* being studied scientifically, particularly in terms of its relationship to the brain, most of its scientist-students would agree that we do not understand it.

Can we nevertheless define some possible requirements for consciousness? Each of the following propositions can be and has been disagreed with, but they are not unreasonable. First, for there to be consciousness, there must be an entity that is aware, distinct from what it is aware of; and conversely, it must be in some kind of communication with what it is aware of. Second, such an entity must have sufficient complexity to both maintain itself—for a time—as an entity, and to apprehend externals. In us, the senses and the brain that analyses them fulfill the latter function and contribute to the former. Third, the entity must be capable of change, growth, or learning; therefore it must have a memory.

The argument can be made that a computer can fulfill all these requirements, and there are many who argue that computers are or can become conscious, which seems intuitively unlikely to many of us. What is missing in a computer? Perhaps it is the awareness of awareness. We can be aware of ourselves being aware of the world; indeed, degrees or levels of consciousness can be measured in terms of the amount of awareness and concomitant self-awareness. Being aware of its awareness seems an unlikely possibility for a computer. Or perhaps it has to do with the sense of meaningfulness: consciousness requires an emotional component. Or perhaps there is a non-algorithmic element to our consciousness, as postulated by Penrose.[1] Or are these three different aspects of the same thing?

Clearly, self-awareness requires the existence of an entity with at least the characteristics described earlier.

Another approach to the question of consciousness is to inquire into its dimensionality. The inner world of awareness is not located in ordinary space or time. One can argue that it is located in the brain, but it cannot be found there, neither by neurosurgeons, nor by MRI scans nor electroencephalograms; all that can be found are tissues and cells and electrical activities that we know are related to inner events, but these are not the inner life itself. Similarly, although the inner life proceeds in time, and perhaps even gives rise to the sense of time, its relationship to time is unlike that of outer world phenomena. The inner world moves effortlessly between past and future, and between actuality and possibility.

However, as the outer world has been probed more and more deeply by physics, it also turns out to be made of elements that defy our ordinary conceptions of space and time—concepts which we acquire through our relationships with macroscopic objects and events. The study of light, or electromagnetic radiation, has led to the paradoxical findings of relativity and quantum physics.[2] Light has no mass, and although it has a fixed speed, for itself there is no spacetime interval between its emission and reception, according to Einstein's theory. A radio or television program, emitted from the station transmitter and going off in all directions through space, can be considered an entity, but its entityness is quite different from that of a table, or a dog, or a human being. It can be picked up by a receiver on the moon, or Mars, or a space vehicle—it seems to multiply itself miraculously, like the loaves and fishes of the Gospels. Any number of such transmissions co-exist right around my head, requiring only an appropriately tuned receiver to become manifest: they do not take up any ordinary space.

Although there is no agreement as to the neural basis of consciousness, there is increasing evidence that it may be related to coherent electromagnetic activity in our brains.[3] Several scientists have argued that the entire body has an electromagnetic shape or coherence.[4] It may be that the level of consciousness is related to the degree of coherence, or harmoniousness, of the electromagnetic vibrations of the different parts of the brain, or even of the whole body. If so, then the unusual dimensionality of the inner life may be of the same nature as that of light.

Does that mean that a radio or television program is a conscious entity? One would think not: it does not fulfill some of the criteria postulated above. It is not open to change or growth, once emitted; it is not in a semipermeable give and take relationship with the world outside itself. So a conscious entity, if it has an electromagnetic character, must have additional elements.

We are not going to solve these issues here; we have only arrived at hints into the mysterious nature of consciousness. With these hints we can now resume the initial inquiry.

The earth can no longer be thought of as simply a large spherical mass whose only interaction with the sun, other planets, and the universe beyond the solar system is a gravitational one. Nowadays the earth is known to have an electromagnetic body, the magnetosphere, which interacts with the solar wind, the stream of particles emanating in all directions from the sun, in very complex ways. The aurora borealis is one of the most spectacular manifestations of this interaction, which also causes disruptions of our radio transmissions, affects the weather, and generally manifests itself as a dynamic, constantly changing relationship. In turn, the magnetosphere, elongated in the direction opposite the sun by the solar wind, extends beyond the moon's orbit, and the moon passes through the earth's magnetic "tail" when it is full. The solar system as a whole has an electromagnetic body of its own, which in turn interacts with the electromagnetic emanations of the galaxy. The galaxy itself has such a structure, relating it to the larger universe.

Perhaps the most important feature of living organisms is that they have mechanisms for utilizing electrical potential energy. A cell has a semipermeable membrane, allowing in some substances and not others, which defines it as an entity. The selectivity of the membrane extends to ions, so that there is a separation of charge across the membrane; the electrical energy so generated is used to power many of the functions of the cell. Nerve cells use this mechanism to generate the impulses that travel from neuron to neuron, enabling the complex ever-changing interactions that ultimately give rise to our perceptions, actions, and awareness. Similarly, the earth's magnetosphere serves as a selective partial barrier, deflecting some of the particles of the solar wind and driving others into specific channels along lines of force. These mechanisms create charge separation and electrical potential energy, in turn giving rise to electrical currents, high above the earth. These phenomena have been known for only a relatively short time, and their purposes are largely still unclear.

The sun is known to have many complex interrelated vibrational modes, from the sunspot cycle to much more rapid vibrations. These vibrations are transmitted to the earth's magnetosphere and influence processes on earth.

The earth then has a semipermeable "membrane," which allows in some influences and deflects others. This makes its structure analogous to that of an organism, or a cell. In Gurdjieff's terms, it is a "cosmos." [5] A cosmos could be defined as a relatively independent, self-sustaining, living entity, which has a selective relationship with the outside world. All cosmoses contain different levels of materiality. We contain a skeleton, muscles and organs, tubes conveying fluids from one place to another, a nervous system controlling the whole. We are made of solids, liquids, gases, electromagnetic energies, and perhaps other materialities we are as yet unaware of. In general the more physical aspects of an organism are of coarser materialities, and its "life" and awareness of progressively finer ones, each interacting with the others in complex ways. At each level there are interactions with the environment. We are tied to the earth by gravity, as the earth is tied to the sun. We take in food, water, air, and sense impressions; we emit wastes that are food for bacteria, altered air that is food for plants, manifestations that serve as impressions for other organisms. We think of our body as the contents of the skin, but the body of the inner life contains all that we are aware of. The earth also interacts with the universe on many levels. If our consciousness is related to a dynamic, ever-changing array of electromagnetic vibrations, interacting with the environment, perhaps the earth too is conscious, as it has all the necessary elements and relationships.

A larger cosmos is composed of smaller ones. We are made of cells—"microcosmoses." To our ordinary perception, most of our cells are just beyond the limit of visibility. We are not aware of them as individuals. The brain is thought to be composed of 100 billion nerve cells. Their combined activity gives rise to our perceptions. Many billions of organisms inhabit the earth's surface. Billions of stars compose the galaxy, billions of galaxies make up the universe. Does our consciousness contribute to the earth's in a similar way that each of our neurons contributes to ours?

Each of us is then as expendable for the earth as any one of our cells is expendable for our body or brain. Yet there is another level of interaction. For I am not just my body, I am everything I am aware of: the galaxies, the earth, other beings, and all of their histories. In a living, conscious universe, the small and the large are related in an unfathomable way.

Smaller than the small, greater than the great, the Self is set in the heart of every creature.

Knowing that the individual Self, eater of the fruit of action, is the universal Self, maker of past and future, the wise man knows he has nothing to fear.

Born in the beginning from meditation, born from the waters, having entered the secret place of the heart, He looks forth through beings. That is Self.[6]

Is the earth conscious? Is the sun conscious, or rather its larger body, the solar system? The galaxy? The universe? These are not questions that can be answered scientifically, at least not now, and perhaps never. We know ourselves to be conscious, but that is not understood scientifically either. Nevertheless, the features that seem to be necessary for consciousness may be present in each of these cosmoses.

NOTES

1. Roger Penrose, *The Emperor's New Mind*. Oxford University Press, 1989. A process is algorithmic if it can be described by a sequence (even a very long one) of logical or mathematical steps, and therefore can be performed by a machine.

2. From "The Eye of the Beholder" in this volume.

3. Gerald M. Edelman and Giulio Tononi, *A Universe of Consciousness: How Matter Becomes Imagination* (Basic Books, 2000).

4. Robert O. Becker, M.D. and Gary Selden, *The Body Electric: Electromagnetism and the Foundation of Life* (Quill, William Murrow, New York, 1985).

5. P.D. Ouspensky, *In Search of the Miraculous: Fragments of an Unknown Teaching* (Harcourt, Brace & World, Inc., New York, 1949).

6. *Katha Upanishad*, edited by the author based on multiple translations.

LAWS, MIRACLES, AND SCIENCE

The term "miracle" has several different meanings, all related but partially contradictory. In common usage, a miracle can be simply an unlikely event: "It's a miracle I managed to get here through all that traffic." Less mundane is the phrase "miracles of modern science," the miracle being the result of an understanding and manipulation of the world at levels beyond our natural visual and tactile comprehension. For the religious, a miracle implies an intervention by God or his representatives in which the laws we are subject to are suspended: a miracle must not be explicable by any law of nature. But this leads to a philosophical conundrum: why would the Deity counteract his own laws, oppose his own will as expressed in those laws, contradicting the very meaning of law? If a law can be suspended, it is not an absolute law. Thus for the scientist, the idea of a miracle is absurd: such exceptions to law require that either the exceptions be explained away, or the law be made more accurate—or supplanted entirely. However, it is this evidence of a higher power, above the laws of nature, which gives the miraculous its appeal, and particularly if that higher power interacts directly with us, shows direct concern for us—we who are otherwise rendered insignificant in the face of the immense unyielding forces of nature. The entire story of Christ is of miraculous intercession on our behalf.

> For the law was given by Moses, *but* grace and truth came by Jesus Christ. (John 1:17)

A view of the miraculous which neither defies logic, nor denies the existence of higher levels to which we can be attuned, is expressed by Gurdjieff:

> The manifestations of the laws of one cosmos in another cosmos constitutes what we call a *miracle*. A miracle is not a breaking of laws, nor is it a phenomenon outside laws. It is a phenomenon which takes place according to the laws of another cosmos. These laws are incomprehensible and unknown to us, and are therefore *miraculous*.[1]

This definition of the miraculous is worth exploring, because it suggests the possibility of a reconciliation between the scientific and religious approaches to understanding, a reconciliation which seems to be one of the main intellectual enterprises of our time. But the stumbling block, for a scientist, has now shifted to the idea of cosmoses. In the modern view, there is but one cosmos, subject to universal laws. While there are degrees of complexity, everything is fundamentally made of the same stuff, and is on the same level. Religious and mystical cosmologies contain at least two levels (often as many as seven).

> Except a man be born of water and of the Spirit, he cannot enter into the kingdom of God. That which is born of the flesh is flesh; and that which is born of the Spirit is spirit. (John 3:5-6)

What is meant by this ancient idea of cosmoses, or of levels? These correspond to levels of materiality, levels of laws, levels of consciousness. According to Gurdjieff, as one descends from God to the minerals of the earth, one encounters progressively denser matter, slower vibrations, less freedom, and less consciousness. But the cosmoses are not physically separate entities: the materials of the different cosmoses interpenetrate one another, as a solid can be saturated with liquid and a liquid with gas, and as a man can be filled with the Spirit.

Does this make any sense in modern terms? The states of matter known to us—solid, liquid, gaseous, plasma, radiant energy—certainly correspond to different kinds of materiality and obey different laws, have different degrees of freedom. The law of gravity may apply to everything, but a rock, a stream, a wind, and the light from the sun are not affected by it in the same way. Were a rock to move almost instantaneously from one place to another, as light does, this would constitute a miracle. An ordinary man cannot walk on water, but the wind can. Many of the classic miracles consist of substances undergoing transformations which are ordinarily forbidden to them, but are not forbidden to substances of a different materiality. Loaves and fishes cannot multiply rapidly, but shadows can if we turn on more lights, and sounds can if there are sufficient echoing surfaces. We cannot see through suitcases or flesh, but x-rays do, providing one of the miracles of modern science.

Another attribute of cosmoses is: "one cosmos is related to another as zero to infinity."[2] This is also the definition of a dimension: an infinite number of points fit into a line, an infinite number of lines into a plane, etc. Certainly phenomena from a higher dimensional world would appear miraculous to an inhabitant of a lower dimensional one, such as the often-cited example of a hand going through Flatland. A hand plunging into water, to a hypothetical observer on the surface of the water, who knew of no other spatial dimensions than the two of that surface, would seem a miraculous phenomenon: five round entities appear out of nowhere and then merge to become a larger entity. Any number of other inexplicable (to a Flatlander) happenings can be imagined, having in common two characteristics: effects without perceptible causes, and connectedness between things which appear separate.

But these characteristics commonly appear in our world as well: a magnet creating an orderly pattern out of a randomly strewn bunch of iron filings, or the connection between a radio and the broadcasting station. These depend on invisible electromagnetic forces and vibrations, which we do not generally interpret as either miraculous or other-dimensional. Yet the scientific description of electromagnetism and other quantum phenomena requires the use of complex numbers (utilizing the square root of minus one), to describe probability amplitudes. This adds two mathematical dimensions to the familiar three of space and one of time. The properties of electromagnetic waves, as described by the theories of relativity and quantum mechanics, defy our Euclidian notions of space and time. Perhaps these vibrations should be regarded as belonging to another dimension, or level, or cosmos.

> The wind bloweth where it listeth, and thou hearest the sound thereof, but canst not tell whence it cometh, and whither it goeth: so is every one that is born of the Spirit. (John 3: 8)

Another characteristic of miracles is that they involve the unlikely creation of order. Restoration of life to a dead person, or of health or sight to a sick one, or the materialization of things out of thin air, are all examples. Order is being referred to even in the commonplace usage of the word miracle: getting somewhere or accomplishing something against all odds. It is also what is striking in the sudden rearrangement of iron filings into a pattern by a magnetic field. And it is the incredibly complex order that characterizes a living organism, despite the overwhelming and ever-present forces of dispersal—entropy—which prompts the expression "the miracle of life."

It is evidence of order everywhere, despite entropy, that makes many of us reject the "scientific" notion that it all came about "by accident," particles bumping into each other to create complex molecules, which then somehow arranged themselves into cells, which aggregated into organisms, which in turn evolved by natural selection of random favorable mutations into us. In ancient traditions order comes from above: the sun is a god, who gives life to the earth. Certainly this is so: the energy from the sun, electromagnetic energy, which can be regarded as coming from a higher level, interacts with the materials of the earth, creating and sustaining life. Beyond the sun, we can feel that there must be a Creator, or at least a primordial creative force. Even the scientific atheist requires creative forces, namely the fundamental laws of the universe: gravity, electromagnetism, and the nuclear forces, as well as the array of subatomic particles that make up everything. It is the particularities of these fundamental elements that makes possible the incredible complexity of the universe. And these seem increasingly to be best represented by the basic mathematical patterns of group theory. Who created the mathematical patterns? It is almost like asking what came before the beginning, a matter of pure belief without possible evidential resolution.

Or perhaps belief itself, or more correctly, *faith*, is a perceptual capacity which can see the divine order in the universe. Faith, in the gospels, can be seen as referring to a heightened emotional perception, which needs to be developed in us. In a state of heightened emotion, the connectedness of everything, the life in everything, can be directly perceived. This perception is accompanied by a sense of wonder, of awe. This is the true miracle:

> For verily I say unto you, If ye have faith as a grain of mustard seed, ye shall say unto this mountain, Remove hence to yonder place; and it shall remove; and nothing shall be impossible unto you. (Matthew 17:20)

Referring to love, which with faith and hope are the three aspects of a higher emotional perceptual capacity, Paul says:

> For now we see through a glass darkly, but then face to face. (1 Corinthians 13:12)

The miracles of the gospels can thus be interpreted allegorically, while losing none of their miraculous nature; they can be regarded as parables, referring to an inner rather than an outer transformation.[3] Many of the key words in the gospels have multiple meanings, often stated rather explicitly. After making the blind man see, Jesus says:

> For judgment I am come into this world, that they which see not might see; and that they which see might be made blind. And some of the Pharisees which were with him heard these words and said unto him, Are we blind also? Jesus said unto them, If ye were blind, ye should have no sin; but now ye say, We see; therefore your sin remaineth. (John 9:39-41)

And:

> Except a man be born again, he cannot see the kingdom of God. Nicodemus saith unto him, How can a man be born when he is old? Can he enter the second time into his mother's womb, and be born? Jesus answered, Verily, verily, I say unto thee, Except a man be born of water and of the Spirit, he cannot enter into the kingdom of God. (John 3:3-6)

In us, many levels of materiality exist. Our solid skeletons, subject to mechanical forces, move about the joints, pulled by muscles which require nutrients dissolved in our bodily fluids. Oxygen, also carried in those fluids, and available because of the interaction of sunlight with plants, provides energy for the muscles and other organs. The energy of life is fundamentally electric in nature, made of potential differences across cellular membranes, and stored in molecules which release their energy by assuming less energetic configurations of their electrons in relation to their atomic nuclei. The nervous system operates by means of electromagnetic signals. Our thoughts are not restricted in space and time in the same way our bodies are. And beyond that, somehow consciousness arises, in relation to complex electromagnetic patterns. It is miraculous that life can animate our flesh, and that consciousness can illuminate our life. The miracles of the Gospels refer to a further miracle: the development of a higher consciousness which can overcome the ultimate human constraint, death.

NOTES

1. P.D. Ouspensky, *In Search of the Miraculous: Fragments of an Unknown Teaching* (New York: Harcourt, Brace & World, 1949), pp. 207–208.

2. *Ibid*, p. 206.

3. Maurice Nicoll, *The New Man: An Interpretation of Some Parables and Miracles of Christ* (London: Vincent Stuart, 1961). See also Ravi Ravindra, *The Yoga of the Christ in the Gospel According to St. John* (Rockport, Mass.: Element Books, 1990).

The Materiality
of the Soul

The existence of a soul that survives the death of the physical body is a tenet of most religious teachings, and it is likely that a majority of human beings believes in a soul, in one form or another. Modern science, however, despite its impressive accomplishments in elucidating the workings of the world, has not shown that the soul exists; many scientists would probably say that the weight of scientific evidence is against its existence. This is one of the major factors that maintain the divide between science and religion. For some, either science or religion is simply not true. For others, such as some scientists who hold religious beliefs, both are true, but refer to completely separate realms of truth about the world. This is an illogical dichotomy, and to hold it implies a serious limitation of human reason. Another option, accepting that there is one world and that all truths about it must be related, is to say that science simply has not achieved the development and subtlety necessary to encompass spiritual questions, such as the existence of God or of the soul. If one takes this route, then one can at least begin to frame relevant questions.

If there is a soul that survives death, then, for its existence to be compatible with a scientific world view, it must logically have certain characteristics. Although some of the following will engender disagreement, a reasonable list of such characteristics might include:

1. The soul must be an entity, that is an organized arrangement of some kind of material substance, which can maintain its integrity for a time or perhaps forever.

2. Its materiality must be much finer and lighter than that of the physical body.

3. It must be conscious and self-aware.

4. It must not be completely isolated, that is, it must have some kind of perception of the world outside of it, and probably some means of action.

5. For the previous two characteristics to be possible, it must have a significant degree of organizational complexity.

There are many kinds of entities, if simply defined as above. Organisms are clearly entities, and probably first come to mind as exemplifying the concept. But a book is also an entity, as is a radio or television program. Each is a carrier of an organized set of ideas, though of quite different materialities. A book is a physical object, but the materiality of a radio or television program is quite different, consisting of organized electromagnetic wave patterns, which travel from their source in all directions at the speed of light. The materiality of electromagnetic waves is quite different from that of ordinary objects. Their relationship to space and time is different, as elucidated by the theories of relativity and quantum mechanics, and leads to many paradoxical properties, paradoxical for our brains, which are trained by the objects in our world. For instance, electromagnetic waves, also described as photons, have no rest mass, cannot be at rest, and any number of them can occupy the same space without getting in each other's way.

For that matter, an idea, or set of ideas, in themselves could be considered entities, in that they have a maintained integrity over time. But can they be considered to have any materiality of their own, aside from that of their carriers—books, electromagnetic waves, brains? If a person's entire stock of memories, skills, perceptual abilities, etc., were reproduced in a computer, would this constitute a soul? Most scientists would probably argue that the "software," in the abstract, has no material existence. In any case a computer "soul" is nowhere near yet possible (if it ever will be), so is not decisive with regard to the question of whether we have souls. Still, for some, the idea of living on in other people's memories, or in one's works, is the only, and a meaningful, sense in which we continue to live after death.

If we accept that the soul must have some kind of materiality, the next question might be: Is it material of a sort as of yet completely unknown to us, or could it be a materiality we are familiar with, but have not considered in this way? To speculate further, we might adopt the latter hypothesis.

Could the soul be something like a radio or television program, an entity of a different and lighter materiality, with a different relationship to space and time? Our bodies certainly contain organized electromagnetic patterns of all sorts, and our consciousness may fundamentally be related to such patterns, generated in our brains.[1] For that matter, we are continually emitting weak electromagnetic waves, which can be picked up by electroencephalography and magnetoencephalography. These wave patterns are continually changing, in tandem with our changing states and contents of consciousness. But once emitted, a radio or television program no longer changes, except to dissipate as it expands in space, unless it is modified by another entity. Though it travels at the speed of light in all directions, it is static in its organizational pattern. The same would be true of the electromagnetic wave patterns emitted by our bodies. They would not fulfill the criterion of being in an ongoing relationship to the world outside themselves, nor could they be self aware in such a relationship.

The existence of separate organisms and their ability to metabolize—to selectively take in and transform substances from the outside so as to maintain themselves and develop—depends primarily on the existence and transformation of electromagnetic patterns. The metabolism of food is essentially a sequence of changes in the electromagnetic configurations of molecules. All cells depend for their functioning on a potential difference across the cell membrane. Nerve cells communicate by virtue of transient changes in these potentials; the brain is in essence a vastly complex pattern of constantly changing electromagnetic interactions.

Since life and consciousness both seem to depend critically on complex electromagnetic interactions, is there a material that is finer and lighter than our solid and liquid bodies that can support such interactions, and fulfill the criteria enumerated above? An ordinary gas does not have enough potential complexity, and is electrically neutral, but a plasma has all the needed possibilities. A plasma is like a gas but is sufficiently ionized—the atoms have been partially stripped of their electrons, so that it contains free electrons and positively charged nuclei—that it can manifest extremely complex behavior. It becomes electrically conductive and responsive to electric and magnetic fields, both those generated within itself and those outside. Filaments, membranes, and cells can form, related to the separation of charges. Simple plasmas include fluorescent lights and lightning. The sun is a very complex plasma. The solar system is permeated with plasmas: the solar wind, the earth's magnetosphere and ionosphere, and the aurora borealis are all complex interrelated plasmas. In fact, plasma is the most common state of matter in the universe, connecting all of its constituents. One of the pioneers of plasma physics, Hannes Alfven, said:

> From the cosmological point of view, the most important new space research discovery is probably the cellular structure of space. As has been seen in every region of space accessible to *in situ* measurements, there are a number of "cell walls," sheets of electric currents, which divide space into compartments with different magnetization, temperature, density, etc.[2]

If the universe is made of cells, perhaps we, being made of cells, are truly made in God's image.

So the plasma, the fourth state of matter, could fulfill the criteria needed to be the material of which the soul is composed. It can form entities, is lighter and finer than the body, can interact with its surroundings, and can be of sufficient complexity to support consciousness. Since consciousness supported by the brain seems to depend critically on electromagnetic interactions, perhaps it is even more suited than the brain to support consciousness.

G.I. Gurdjieff taught that there are several levels, or stages of development, of the soul, the first corresponding to what has been called the astral body, which he also called the *"kesdjan* body": "The matter of this astral body, in its vibrations, corresponds to matter of the sun's emanations and is, theoretically, indestructible within the confines of the earth and its atmosphere."[3]

One of Gurdjieff's interesting and radical ideas is that the last supper was a ritual intended to establish a connection between Jesus and his disciples that would survive Jesus' death. The bread and wine were real flesh and blood, and their consumption by the disciples enabled a threadlike connection to be established between Jesus' soul and the souls of the disciples.[4] In another story in his book, *All and Everything*, Gurdjieff describes the attempt by Tibetan monks to establish contact with their suddenly and unexpectedly dead leader. They did this by trying to permeate his dead body with the emanations (hanbledzoin) of their own kesdjan bodies. However, "because the connection with his kesdjan body had not been established beforehand, their hanbledzoin did not accomplish what was expected but only accumulated chaotically over his planetary body." There were violent thunderstorms in the area, and the contact between the lightning and the handbledzoin of the monks resulted in a huge explosion.[5]

There are many such hints in Gurdjieff's teaching, which suggest that at least the astral body's materiality is that of a plasma. The materiality of the other levels of the soul remains more obscure.

One aspect of the progressive development of bodies of finer materiality is a changing relationship to time:

> "Of course," according to physicist Richard Fitzpatrick, "bound systems can display extreme complexity of structure: e.g., a protein molecule. Complexity in a plasma is somewhat different, being expressed *temporally* as much as *spatially*. It is predominantly characterized by the excitation of an enormous variety of *collective* dynamincal modes."[6]

As mentioned above, and discussed in more detail elsewhere, electromagnetic vibrations have a different relationship to time from ordinary matter. In addition, the inner life seems to be located not in space, but in a multidimensional time.[7] A plasma, being in essence an entity in which electromagnetic vibrations make up its very structure, has a different relationship to time.

As A.R. Orage, one of Gurdjieff's prominent pupils, put it,

> Time is a three-fold stream passing through our three centers (mind, emotions, instinct). We fish in that stream. What we catch is ours. What we don't catch is gone. Time does not wait for us to catch everything in that stream. If we catch enough, we have enough to create the three bodies and become enduring.

Time is the sum of our potential experiences, the totality of our possible experiences. We live our experiences successively. Succession is the first dimension of time. To be able to live experiences *simultaneously* is adding a second dimension to time. To be aware of this simultaneity is solid time. When we have identified ourselves with time, as Revelation says, "Time shall be no longer."[8]

Many religions believe in the reincarnation of souls. How could this be possible if a soul is an organism made of plasma? To speculate even more wildly, studies of DNA indicate that it has unique electromagnetic properties because of its helical structure, which permits the core of the DNA helix to be highly conductive to electrons, making DNA very sensitive to electromagnetic fields in a wide range of frequencies.[9] It is conceivable that the soul could thus interact with the DNA of an embryo to specify many of the embryo's characteristics, by selectively turning on the transcription of some proteins versus others, thereby adding to the effects of heredity and environment. It is well known that only a small fraction of human DNA (genes) actually codes for proteins, and that non-coding DNA includes portions that are involved in regulating the expression of the coding DNA. The reception of electromagnetic signals may be one of the roles of this part of the DNA.

For some scientists, any talk of a soul is nonsense; there cannot be such a thing in a material world. For some spiritual purists, the realm of the spirit is unapproachable by science; it is of a different, nonmaterial, order. But Gurdjieff famously said that everything is material:

Everything in this universe can be weighed and measured. The absolute is as material, as weighable and measurable, as the moon, or as man. If the absolute is God, it means that God can be weighed and measured, resolved into component elements, "calculated," and expressed in the form of a definite formula.[10]

If there is a soul, it seems, it must conform to universal laws.

NOTES

1. Christian Wertenbaker, "The Light of the Beholder," in this volume.

2. Wikipedia, Plasma Physics.

3. G.I. Gurdjieff, *Views from the Real World* (New York, E.P. Dutton & Co., Inc., 1973), p. 97.

4. P.D. Ouspensky, *In Search of the Miraculous: Fragments of an Unknown Teaching* (New

York, Harcourt, Brace & World, Inc., 1949), p. 207–208.

5. G.I. Gurdjieff, *Beelzebub's Tales to his Grandson. All and Everything, First Series* (New York, Jeremy P. Tarcher/Penguin, 1992), p. 669.

6. Richard Fitzpatrick, internet course on plasma physics.

7. Christian Wertenbaker, "One and One make One," in this volume.

8. A.R. Orage, quoted by Louise Welch, in *Orage with Gurdjieff in America* (Boston, London, Melbourne and Henley, Routledge & Kegan Paul, 1982), pp. 79–80.

9. M. Blank, R. Goodman: "DNA is a fractal antenna in electromagnetic fields." Int J Radiat Biol 2011 Apr; 87(4):409–15.

10. P.D. Ouspensky, p. 86.

5

THE ROLE OF MAN
IN THE COSMOS

THE FULLNESS OF THE VOID

In spiritual and mystical traditions, thought is often regarded as an impediment to the direct perception of reality and of God:

> Which of you by taking thought can add one cubit unto his stature? And why take ye thought for rainment? Consider the lilies of the field, how they grow; they toil not, neither do they spin. (Matthew 6:27-28)

> The Tao that can be told is not the eternal Tao.
> The name that can be named is not the eternal name.
> The nameless is the beginning of heaven and earth.
> The named is the mother of ten thousand things.
> Ever desireless, one can see the mystery.
> Ever desiring, one can see the manifestations.[1]

In more secular, and especially in scientific circles, the highly developed human capacity for thought—for conceptualization, abstraction, logic, and language—is seen as the pinnacle of the evolutionary process. The mind has also been considered a manifestation of the most holy:

> Let each man think himself an act of God.
> His mind a thought, his life a breath of God.[2]

Are these views truly opposed, or is more examination needed?

A little introspection reveals that thought is not always the same. Daydreaming is an almost random succession of words and images, one leading to another by some associative link. What we usually call thinking, an attempt to figure out a problem, involves a seemingly more deliberate sequence of associations, but often interrupted by distractions of various sorts. Then there are moments of sudden comprehension—"I've got it!"—of a concept that has eluded us, or even of a whole collection of concepts, when in an instant something is grasped that might take hours to explain fully. These moments can appear after long tedious deliberation that seems to go nowhere, or spontaneously. Extreme examples of this—Mozart reportedly could hear an entire symphony in his head all at once—are experienced by only a few. However, our thought clearly has many different modes of operation, more than these few examples illustrate, and we often shift from one to another and experience mixtures of modes.

What distinguishes thought from other physical and mental capacities? In Indian philosophy, thought is regarded as a sense organ, although on a somewhat higher level than the five senses, a coordinator of the senses and actions.

Five senses of perception—hearing, touch, sight, taste, and smell; five senses of action—speech, sensory apprehension [grasp], locomotion, excretion, generation; and the lower level of thought that coordinates these senses as a whole (manas).[3]

We define the senses by their range of perception. Different modalities of touch respond to heat, cold, sharpness, or light touch; the eyes are sensitive to patterns of light, darkness, and color. What does thought apprehend? Thought can be said to perceive the *form* of things, at varying levels of abstraction. Supplied with raw data by the senses, we perceive the shape of objects, and are able to see them as a whole in the mind's eye, despite different sensory perspectives. At a more abstract level we conceive of classes of things— tables, flowers, etc. Still more abstract are concepts like the number three, or multiplication, or causality, or empathy. No longer tied to objects, these concepts can be combined—we can multiply three by seven, or conceive that empathy caused us to act in a certain way. By means of complex abstractions we can derive laws of nature and predict the results of combinations of phenomena, so these abstractions must be as real as mountains and flowers— perhaps more so, being more fundamental and enduring.

G.I. Gurdjieff, the Greek-Armenian philosopher and mystic, once famously said that everything in the universe is material, and that even God can be weighed and measured.[4] But the materiality of different kinds of things is vastly different, as indicated by their properties. At the level of ordinary matter, solids, liquids, and gases have different degrees of freedom with regard to the three dimensions of space. Electromagnetic waves, like light, have properties so different that the concepts of space and time had to be modified in order to accommodate them.[5]

Ideas, the things of thought, have miraculous properties compared to grosser material things. Like the loaves and fishes of the gospels, they can proliferate with little effort, occupying millions of minds without apparent attenuation. In fact, they share many of the properties of light and other electromagnetic waves. They move at similar speeds. A television program, a bundle of electromagnetic waves, occupies many television sets simultaneously. For that matter, a broadcast can be regarded as a little packet of ideas being mailed all over the world at the speed of light, yet needing a proper receiving apparatus—not just the radio or television set, but the minds that listen or watch—to become alive, just as instant coffee needs hot water.

Thought, then, allows us to sense a level of reality more subtle than that of ordinary material things, and perhaps we humans are uniquely gifted in this regard. The capacity for abstract thought in animals, even in our closest relatives, the chimpanzees, seems clearly to be much less developed, so much so as to be of a different order altogether.

The lower levels of this capacity to abstract are common to higher animals and humans, and have been partially mapped to the brain. Thus, over thirty cerebral cortical areas devoted to vision are known, many of which specialize in analyzing a particular aspect of the visual scene, by abstracting a useful subset of information from the raw visual data. For instance, one area provides for color constancy. Different illuminations, such as tungsten bulbs versus daylight, change the wavelengths of light emitted by objects drastically. A photograph taken with daylight film in a room illuminated by a tungsten bulb looks very yellow. However, when *we* move from one illumination to another we notice little change in the colors of objects. This "color constancy" is provided by a particular visual area in the brain, which has to compute the "true" colors of objects by filtering out the diffuse bias produced by the illuminating wavelengths of light. This involves a large number of comparisons that occur simultaneously, in parallel. The cerebrum is specialized for this kind of computation, with over 100 billion nerve cells so massively interconnected that generally a maximum of five steps will get from any randomly chosen neuron to any other. Other areas are specialized to recognize particular objects regardless of their visual size or angle of viewing. And so on. Ultimately our brains abstract concepts, designated by words: table, chair, number, algebra, kindness, universe.

The central mystery of neuroscience, and a subject much debated today, is where, or how, or even why, conscious awareness comes into this process. Clearly much of this computational process can go on without conscious awareness, or sometimes with and sometimes without it. We are certainly not aware of the activity of individual neurons, or of much of the lower-level computations that allow us to move a limb precisely to catch a ball, produce words and sentences, or recognize objects at a glance. And everyone has had the experience of being engaged in conversation while walking or driving, and realizing that one has successfully stayed on course, avoided obstacles, etc., without any apparent awareness or memory of the complex sensorimotor interactions that have been taking place. We are able to massively multitask while paying attention to only a small subset of what is going on.

These three functions, of conscious awareness, memory, and attention, would appear to be very interrelated. The small subset of brain processes of which we are aware seems to depend on attention, and these same processes are more likely to be laid down in memory. However some degree of attention can exist without apparent consciousness, and in certain neurological disorders a person can be conscious without any ability to lay down long-term memories. So what is the critical element that determines the degree of consciousness? Antonio Damasio, a neurologist and neuroscientist, argues in his book *The Feeling of What Happens: Body and Emotion in the Making of*

Consciousness that it is the conjunction of self-awareness and awareness of the external world at the same time. "Consciousness, as we commonly think of it, from its basic levels to its most complex, is the unified mental pattern that brings together the object and the self."[6]

While Damasio regards all forms of consciousness as depending on this conjunction, in some spiritual traditions self-awareness is regarded as something that is not always present, has varying degrees of intensity, and needs to be developed. G.I. Gurdjieff distinguished between an automatic sort of awareness, in which we spend much of our lives, and true consciousness, which begins with self-consciousness, a state he called *self-remembering*.

> In ordinary conditions of life we do not *remember ourselves*; we do not remember, that is, we do not feel ourselves, we are not aware of ourselves at the moment of a perception, of an emotion, of a thought or of an action. If a man understands this and tries to remember himself, every impression he receives while remembering himself will, so to speak, be doubled.[7]

This emphasis on self-awareness as the gateway to a higher level of consciousness is also described in Buddhist texts:

> Bhikkhus!
> On another aspect of this practice,
> a person
> knows
> when going,
> I am going
> when standing,
> I am standing
> when sitting,
> I am sitting
> and when lying down,
> I am lying down.
> Thus again and again
> his attention is directed
> on the body,
> and again and again
> he knows
> the body
> in all of its
> movements and postures.[8]

A key realization is that the degree of consciousness fluctuates constantly. The illusion of constancy arises in part because simply asking the question

"Am I conscious?" tends to awaken consciousness for a moment.

> By observing in yourselves the appearance and the disappearance of consciousness you will inevitably see one fact which you neither see nor acknowledge now, and that is that moments of consciousness are very short and are separated by long intervals of completely unconscious, mechanical working of the machine. You will see that you can think, feel, act, speak, work, *without being conscious of it.*[9]

It becomes clear that consciousness, although related, is not the same as thinking, and that one can be variably conscious of one's thoughts. One can observe them just as one observes one's sensations, although it is more difficult. And a deliberate sensing of the body can provide a viewpoint, as it were, from which to observe thought.

One the fundamental aspects of Gurdjieff's psychological ideas is that the human nervous system is composed of three "brains," an intellectual brain, an emotional brain, and a moving-instinctive brain. A similar concept is emphasized by neurologist Paul MacLean, who calls them the "reptilian," the "paleomammalian," and the "neomammalian" brains.[10] Although the functions of these brains are interrelated, self-examination reveals them to be quite distinct in their inner representation. According to Gurdjieff, it is the conjunction of these functions that makes possible a greater degree of consciousness, both of ourselves and of the world around us.

> We must understand that every normal psychic function is a means or an instrument of knowledge. With the help of the mind we see one aspect of things and events, with the help of emotions another aspect, with the help of sensations a third aspect. The most complete knowledge of a given subject possible for us can only be obtained if we examine it simultaneously with our mind, feelings, and sensations. . . . In ordinary conditions man sees the world through a crooked, uneven window.[11]

Perhaps these functions are related to the three "dimensions" of time postulated by P.D. Ouspensky: ordinary time, flowing from past to present to future; a dimension of eternity, or the perpetual existence of every moment; and a dimension of possibilities. For the body clearly is stuck in the present of linear time, moving steadily from past to future, from birth to death. The mind, on the other hand, knows no such boundaries: it moves freely into the past and speculates about the possibilities of the future—even about the possibilities of the past. It lives in the realm of possibilities. And feeling, though more elusive, has a quality of eternity: how I feel now seems that it will be forever; I will always feel this good, or this depressed—suicide attests to this illusion. Only thought, if it is independent enough to not be completely driven by the emotions, can conceive of another possibility, or remember that not so long ago I felt quite differently.[12]

Ordinary associative thought, when divorced from a sense of the body in the present and from the perception of feeling, is one-dimensional, presenting a partial, and therefore distorted, view. The projections of three-dimensional objects into lesser dimensional spaces result in a completely different perception of their nature, as illustrated in Abbott's classic *Flatland*.[13] (For instance, a spherical ball moving through a plane appears to a "plane being" as a point, becoming a growing, then shrinking circle, and finally a point again before disappearing.) When the three perceptual modalities come together, when I feel myself present in the world, perceiving it without distraction, a sense of another dimensionality can appear; everything acquires a new vividness, depth, and meaning. In such a state the mind can appear empty of words and internal images, a "perfect mirror," as the Zen masters would say. Yet the words, and concepts, would be there if I needed them. The possibilities are there also, seen all at once, but there is no need to inventory them. It is as if the droplets—concepts, words, and images—of ordinary thought had vaporized into a three-dimensional timespace, still able to condense if necessary, but now an integrated part of a more whole reality.

> The state where the mind has no impressions of any sort and nothing is beyond its reach (*nirbijah samadhi*) is more intricate than the state of directing the mind towards an object (*Samadhi*).[14]

> In this ultimate state of open availability, the mind instantaneously grasps everything.[15]

> *Consciousness* is a state in which a man knows *all at once* everything that he in general knows...[16]

Most of us only have small partial glimpses of this state, but its existence is attested to by every spiritual tradition.

What could be the neural basis for this perceptual state? This remains a mystery, as even the neural basis for ordinary waking consciousness is unknown. A much lower-level, but probably related, problem is being investigated, however. This is the "binding problem": how does the brain integrate perception? For, as mentioned above, there are some thirty visual areas, to take just one sensory modality, many specialized for analyzing color, shape, location, depth, movement, etc., and no area seems to contain all of the information relevant to a visual percept. There is no downstream spatial location in the brain where everything converges, although the inner experience is of a unified percept. However all the areas are massively interconnected, and a hypothesis is that the electromagnetic brain waves produced by these different areas synchronize when they are related to the same percept—a convergence in time, in the present. There is much evidence to support this, but what it means is controversial. One could postulate, somewhat boldly, that the physical correlate of a more comprehensive consciousness is in fact the integrated electromagnetic activity of the brain, perhaps even of the whole body. When we are distracted, driving a car while drinking coffee, listening to the radio, and worrying about the impression we are going to make on someone, multiple areas of the brain are active and vibrating, but not in any globally synchronized way. The same visual areas that support perception of the outside world are also used for internal imagery; if the screen is playing several movies at once, none will be very clear.

When, however, the mind is a perfect mirror, the entire organism vibrates in harmony, both internally and with the external world. Fully conscious of myself, I become a part of everything.[17]

NOTES

1. Lao Tsu, *Tao Te Ching*, translated by Gia-Fu Feng and Jane English (New York: Vintage Books, Random House, 1997), number one.

2. Philip James Bailey. Festus (1839).

3. B. Bouanchaud, *The Essence of Yoga: Reflections on the Yoga Sutras of Patanjali*, translated from the French by Rosemary Desneux (Portland, Ore.: Rudra Press, 1997), p. 142.

4. P.D. Ouspensky, *In Search of the Miraculous: Fragments of an Unknown Teaching* (New York: Harcourt, 1949), p. 86.

5. "The Eye of the Beholder" in this volume.

6. A.R. Damasio, *The Feeling of What Happens: Body and Emotion in the Making of Consciousness* (New York: Hartcourt, 1999), p. 11.

7. P.D. Ouspensky, *op. cit.*, p. 188.

8. J. Siff, "A Modern Translation of the Satipatthana Sutta," http://meditationproject.com/Siff_Translation.htm.

9. P.D. Ouspensky, *op. cit.*, p. 116–7.

10. P.D. MacLean, *The Triune Brain in Evolution: Role in Paleocerebral Functions.* (New York: Springer, 1990).

11. P.D. Ouspensky, *op. cit.*, pp. 107–8.

12. Although Ouspensky's three time "dimensions" have not been adopted by contemporary science, two time dimensions have been postulated by physicist Stephen Hawking, and the nature of time is mysterious and contradictaory from a scientific point of view. Einstein's relativity theory implies that all of time coexists, the flow of time being an illusion, while quantum theory implies the very real existence of a realm of possibilities, only some of which "collapse" into observed entities. While the existence of two mysteries does not necessarily make them related, it seems likely that the unsolved scientific mysteries of time and of consciousness may well be, and that the current incompatibility of general relativity and quantum mechanics is part of this puzzle.

13. E.A. Abbott, *Flatland: A Romance of Many Dimensions* (Totowa, N. J.: Barnes and Noble, 1963).

14. Yoga Sutra III.8. T.K.V. Desikachar, *Reflections on Yoga Sutras of Patanjali* (Chennai, India: Krishnamacharya Yoga Mandiram, 2004), p. 82.

15. B. Bouanchaud, *op. cit.*, p. 157.

16. P.D. Ouspensky, *op. cit.*, p. 155.

17. Recent research has shown that experienced Buddhist meditators show highly synchronized brain wave activity during meditation, in the same frequency range that is involved in "binding." A, Lutz , L.L. Greischar, N.B. Rawlings, M. Richard, R.J. Davidson, "Long-term meditators self-induce high-amplitude gamma synchrony during mental practice," Proc. Natl. Acad. Sci. U S A. 2004, Nov.16;101(46):16369-73.

THE COSMIC NECESSITY
OF SUFFERING

The question has often been asked: If an all-powerful and benevolent God is watching over us, why is there so much suffering in the world? Thus put, the question makes several assumptions that are probably incorrect. The most evident are that God is watching over us as individuals, or even groups—chosen peoples—and that his power extends to micromanaging our affairs. From a human being's point of view, individual cells, let alone molecules and atoms, are below the threshold of perception, and no particular cell, molecule, or atom is of any interest. We look after our cells by eating, drinking and breathing, but our concern is on a different scale. How much further away from us as individuals must God be: just as our bodies are made up of billions of cells, the universe contains billions of galaxies, each of which is composed of billions of stars, and the earth has billions of people, and many more billions of animals and plants. There is little doubt that the fundamental laws of the physical universe are immutable, and if a God created the world it would have been through the creation of those laws. Why would he break them to intervene in the life of a miniscule component of the creation?

Several possibilities remain: one, there is no God, or higher principle or consciousness from which all creation flows. Second, God is not benevolent, and third, suffering is an inherent and necessary aspect of how the universe works. This third view seems to me the most likely to be correct.

What is suffering? At first one tends to think of physical pain, but suffering is more broadly conceived of as an emotional condition. We suffer because of loss: loss of a loved one, loss of a job or position, loss of capabilities, fear of death. We also suffer greatly from injuries to our egos—loss of perceived appreciation by others. All these sufferings have one thing in common: separation and isolation from the world. The emotional accompaniment of physical pain is also of this nature.

In many religious and mystical traditions the world is a cyclical process: from sin to redemption, from Adam and Eve expelled from the garden of Eden to salvation through Jesus. The goal of Buddhism is liberation from suffering by uniting with a greater universal consciousness. Other traditions, stripped of their more concrete metaphorical trappings, basically describe the same goal: unification with God, with the Beloved, with All. A more detailed and even scientifically plausible elaboration of this idea can be found in the writings of G.I. Gurdjieff, who regarded his view of the world and of humanity's place in it as a restatement of ancient knowledge in a more modern form. In his deliberately obscure major work, *All and Everthing, Beelzebub's Tales to his Grandson*, Gurdjieff states that "His Endlessness"—God—created the universe because his "abode," the "Sun Absolute," was gradually being eroded by time, or entropy.[1] To safeguard the Sun Absolute, His Endlessness changed the fundamental laws governing all processes, which then, of itself, led to the creation of stars, planets, and plants and animals, the ultimate goal of the creation being the evolution of relatively independent conscious beings. If one takes the "abode" of His Endlessness to mean a universal energy of consciousness, then, in Gurdjieff's view, the purpose of the universe is the replenishment and maintenance of this consciousness. Gurdjieff states that the fully developed souls of human beings, and of other similar beings throughout the universe, stand in relation to the whole universe exactly as the individual cells of each of our brains stand in relation to the brain as a whole.[2]

This is the fundamental paradox: just as individual, separate cells, any one of which cannot perceive, think, or cause action by itself, are needed to work together to produce a perceiving, thinking, and acting brain, so consciousness, the ultimate unifier of all things, must arise in individual, separate beings. Thus, as individuals, we have a double duty. First, we must maintain our integrity as separate organisms and the integrity of the population, by finding food and mates and fighting off competitors and predators. The body needs to preserve its separateness from the world. On the other hand, our consciousnesses need to expand to include others and the whole world; awareness rejoins everything together. A collection of perceiving beings, aware of each other and of the world, forms a web of consciousness, thus maintaining the abode of His Endlessness. In some traditions it is put this way: God created the world and man so that He could know Himself. In the Hindu image of Indra's net, the universe is a web of fine filaments; at each intersection is a jewel that reflects all the others. The process of creation is one of individuation and separation; unity is regained when each part reflects and rejoins the whole.

It is quite clear that we are very effective anti-entropy machines on a physical level. Our bodies maintain their complex structure for a long time, and, thanks to our tremendous capacity for abstraction, imagination, foresight, and planning, we maintain our homes, build our cities, and turn piles of raw materials into useful devices. What is less appreciated is that our consciousnesses, fully developed, can perform an analogous function on a higher level.

Consciousness in us is mediated by both feeling and thought, functions of our nervous systems. Those same functions also serve the body and help it maintain its separateness. There is an inherent conflict between these two purposes: selfishness and competition on the one hand, and empathy and oneness with all on the other. As we are, ordinarily, the body's concerns have taken over: our egos, with their endless need for approval and superiority, have usurped our consciousnesses, whose proper higher purpose is the development of compassion and universal awareness. Freud wrote: "The ego is first and foremost a body ego." [3]

This is the human struggle: the liberation of consciousness from its imprisonment by the body's needs. In a biochemical process, chemical bonds must be broken, so that new bonds can be formed to create new molecules. Things must be taken apart before they can be reconstituted into a new form. Similarly, our selfish habits and concerns, side effects of our need to preserve our separate individualities, must give way if we are to unite in a new way with the world. This seeing of the ego and its constructs, the death of illusion, can cause tremendous suffering, ultimately to be replaced by the joy of belonging to a greater consciousness. The self-image the ego has created must be seen through, and must be replaced by a real Self.

In Gurdjieff's terms, the inner growth that leads to the creation of a soul capable of serving universal consciousness depends on the uncovering, under all the layers of ego, of conscience. At one point he writes that conscience arises in human beings from the "emanations of the sorrow" of our creator.[4] One can understand this sorrow as arising from the necessary separation of created beings from the creator, a separation required for the development of individual consciousnesses which ultimately can reunite with Him. In each of us this results in a struggle between the needs of the body and the needs of the soul.

The awakening of conscience necessarily causes suffering. In Gurdjieff's words:

> *Conscience* is a state in which a man *feels all at once* everything that he in general feels, or can feel. And as everyone has within him thousands of contradictory feelings which vary from a deeply hidden realization of his own nothingness and fears of all kinds to the most stupid kind of self-conceit, self-confidence, self-satisfaction, and self-praise, to feel all this *together* would not only be painful but literally unbearable.
>
> If a man whose entire inner world is composed of contradictions were suddenly to feel all these contradictions simultaneously within himself, if he were to feel all at once that he loves everything he hates and hates everything he loves; that he lies when he tells the truth and that he tells the truth when he lies; and if he could feel the shame and horror of it all, this would be the state which is called "conscience."
>
> We may say that there exists the conscience of a man in whom there are no contradictions. This conscience is not suffering; on the contrary it

is joy of a totally new character which we are unable to understand. But even a momentary awakening of conscience in a man who has thousands of different I's is bound to involve suffering. [But] if these moments of conscience become longer and if a man does not fear them but on the contrary co-operates with them and tries to keep and prolong them, an element of very subtle joy, a foretaste of the future "clear consciousness" will gradually enter into these moments.[5]

Separation from the Creator and imprisonment in our selfish bodies seems to be a necessary step in the process of the creation of individual consciousnesses, which then paradoxically can unite with others and with the Creation as a whole, to maintain the "abode" of His Endlessness. As described above, we also become separated from ourselves, because of our failure to face the fundamental dilemma of our existence; we become fractured by contradictory impulses of selfishness and aspiration. Unification with all first requires inner unity.

Does this require there to be illness, famine, war, and devastating earthquakes? Perhaps. Adversity can lead to great inner growth. From a larger perspective, though, many of our sufferings are side effects of the processes by which the universe is maintained. Possibly, if we took our cosmic duty seriously, our suffering could be less random and more appropriate. If we struggled inwardly, suffering our inadequacy in the face of the great purpose offered us, at least murder and war could become unnecessary. We would understand the true nature of suffering—isolation—and work to develop the consciousness that alone can unite us with the Whole.

NOTES

1. G.I. Gurdieff, *Beelzebub's Tales to His Grandson. An Objectively Impartial Criticism of the Life of Man. All and Everything, First Series.* (New York: Jeremy P. Tarcher/Penguin,1992), p. 685.

2. *Ibid.*, p. 712.

3. Sigmund Freud, quoted in Oliver Sacks, *The Man who Mistook his Wife for a Hat* (New York: HarperCollins Publishers, 1990), p. 52.

4. G.I. Gurdieff, *op.cit.*, p. 339.

5. P.D. Ouspensky, *In Search of the Miraculous: Fragments of an Unknown Teaching.* (London, Routledge and Kegan Paul, Ltd., 1950), pp. 155–156.

DOES MAN HAVE THREE BRAINS?
A Tentative Framework for a Theory of Consciousness

The Brain—is wider than the Sky—
For—put them side by side—
The one the other will contain
With ease—and You—beside.

—EMILY DICKINSON

1. The Three Aspects—Dimensions—of Every Thing.

Consider a lamp, for instance. There is one next to me, on the desk. What does it consist of? Suppose we were being questioned by an intelligent English-speaking Martian. We might reply: Well, the base and stem are made of wood, the shade of a metal rim and plastic and cloth covering, the socket of metal and plastic, the bulb of glass, various metals and probably some obscure chemicals, and so on. If pressed, we could elaborate further and explain that the substances we have named are composed of various arrangements and combinations of several atomic elements, which in turn are made of protons, neutrons, and electrons, which in turn...at this point our education might fail us.

But, we are told, so far we have only described an object that occupies a certain amount of space, collects dust, and can get knocked over by the cat. What does it really consist of? Oh, we say, you mean: What does it *do*? Obviously, it provides light for my desk. For that, it has to be plugged into the wall socket, which is connected through miles of wire to a generator somewhere, and the switch has to be on. In fact, all of the parts of the lamp serve this function, without which the lamp doesn't make any sense. The light comes about because the filament in the bulb is heated by the electric alternating current, coming from the generator through the wires, and the heating of the filament causes electrons to be activated in such a way that some frequencies of electromagnetic radiation, light, are emitted...once again, we are wondering about our education.

Where did the lamp come from? It didn't just grow out of the desk, did it? Our interrogator is beginning to sound annoyingly like a four-year-old child. No, my mother-in-law gave it to me for Christmas. She bought it in a store, which got it from the lamp company. The wood came from Sweden, the tungsten from Nigeria, and I'm not sure about the rest. At the company, a man designed it, it was an idea in his head, a plan, which he may also have put on paper, an idea related to the general idea most of us have of "lamp." Then, according to his plan, the wood was cut to this shape, the shade was made like this, and so on.

Although the details of each of these lines of thought elaborate into almost infinite complexity, until we throw up our hands in frustration and declare that we don't really know what a lamp consists of, it does seem clear that without each of these three overall aspects a lamp as we know it cannot exist. A lamp does not exist if the material object is not there; it is only an idea. But it also cannot exist if the idea is not there; without the idea the material could not have taken the form of a lamp. In a sense the idea and the material give each other form, make each other real. And the lamp would not be a lamp if it could not cast light; at best it could be called a dead lamp.

Or take a drum. It also is a material object, the realization of a concept and design, and is missing an essential aspect without the vibrations that animate it. Or a cat. A cat is made of tissues, cells, molecules, atoms. But a design is also immediately evident, although its origin is a source of controversy. The third aspect of a cat is that it is alive. A dead cat is not the same as a living one. This also has been a source of controversy in the past, with legions of scientists searching for the unique energy of life. It is generally agreed now, at least among scientists, that the energies that animate living creatures are the same as those found elsewhere: chemical energy, electrical energy, etc.

Nevertheless, it remains obvious to our direct perception that a dead cat is not a cat at all, and that, regardless of its origin, a cat is a very well designed creature. And of course a cat is a physical object. Taking away any of these three aspects of a cat is inconceivable without completely altering its nature.

In thinking in this way we run into many difficulties, because each of these three views of an object seems to be hopelessly tangled up with the others. The material of the lamp, examined at the subatomic level, becomes as non-solid as the light energy that the lamp produces; in fact, the matter and the energy seem to be made of similar stuff. At that level we also run into the paradoxes of quantum physics, in which everything seems to exist only as a possibility until it is detected, or perceived, much as the ideas in our minds lack substantiality until they take material form. The light of the lamp is also an idea, a concept. The idea of the lamp seems to require a material substrate, our brains, composed of cells, molecules, atoms, etc., and animated by electromagnetic signals similar to those that animate the lamp itself. The three aspects of a thing are not like three separate piles of different materiality that can be examined in isolation. They are more like the three coordinates of a point in three-dimensional space.

A point is not located without all three coordinates; it does not exist in the same way. To claim that the x-coordinate is more real than the others is absurd, though it might seem that way from a restricted viewpoint—for instance, if one is only concerned with the distance between two places along a straight railroad track, only one dimension is important. Similarly, our modern tendency to regard the materiality of an object as more "real" than the life in it or the concept of which it is a manifestation is a restricted view. Furthermore, the three-dimensional coordinate system defining a point can be moved or rotated; all three coordinates of the point then change according to well-known mathematical formulas, but it retains three coordinates. However, each coordinate is now a "blend" of all three of the previous coordinates. This feature lends further support to the idea that the three aspects of things, as described above, are properly thought of as three dimensions, and the confusion caused by the intermingling of categories may begin to become comprehensible.

2. The Three Aspects of our Inner Life

Examination of our inner experience (the only experience we know) also reveals a number of distinguishable components: sensations, perceptions, thoughts, images, memories, emotions, levels of consciousness (sleep, waking), intentions, movements. . . again, however, each of these blends into and is mixed up with the others. Is there a classification of the inner life that seems fundamental?

One of the basic ideas of the cosmology and psychology of G.I. Gurdjieff is that man is a three-brained being, in contrast to other mammals, which are two-brained, and animals like worms and snakes, which are one-brained. He calls the three brains the thinking center, the feeling center, and the moving-instinctive center.[1, 2] This is not unlike the classification of the neurologist Paul MacLean, whose "triune" brain is composed of the "reptilian" brain, controlling basic bodily functions such as breathing, blood pressure, etc., and complex reflex movements; the "paleomammalian" brain, subserving emotion; and the "neomammalian" brain, subserving thought, language, and other intellectual functions.[3]

Such a classification does correspond to our introspective experience. Each of these aspects of brain-function has a different inner quality. Our usual "stream of consciousness" is very much located in the head, a rather haphazard blend of perception, memory and imagination; its idiom is language and image (in all sensory modalities). One of its striking characteristics is its ability to roam without encountering any barriers between past, present, and future, and between actuality and possibility. Without it we could not make plans or use the "retrospectoscope," analyze, or conceptualize. We would have no access to ideas.

Our thought is connected to our movements, but is quite distinct from them. The relationship is like that between executive and workers; thought can give the "moving center" broad instructions, which will be best carried out if the moving center is left alone. We could not walk or talk if we had to think about the contraction of each muscle in sequence, and our performance of newly learned movements tends to be quite clumsy until they become automatic. Our sensations are also quite independent; they generally enter into our "thinking-consciousness" only after extensive filtering and processing, and a constant flow of sensory information is used by the moving center to adjust our balance, avoid obstacles, and modify movements to conform to changes in the environment, all without the participation of thought. The way in which we can drive a car while immersed in conversation or daydreams is a striking example. Finally, instinctive functions seem almost entirely autonomous: thought generally has no direct role in regulating the activities of our intestines, blood vessels, heart, or sweat glands, as they and every other organ carry on a multitude of tasks to maintain the physical plant, with constant adjustments based on need and supplies. We become aware of our instinctive functions when they need to tell us something, like "get some food," or "get rid of the garbage," or "something is wrong," and thought interacts with them only very indirectly. It is like the executive who has no awareness of the boiler room of his factory except through the engineer, by watching the smoke coming out of the stack, or when there's no heat.

The emotions, in some ways, seem the least distinct, the least assignable to a "center," or brain. They seem to be neither as directly linked to our subjective consciousness as are our thoughts, nor as distinct from it as are our bodies' automatic functions. They color both, provide the general "tone" for both body and thought; to a large extent we know them only by their influence on the other two. Our thoughts are guided, or driven, by emotions, whether passion, fear, or simply interest; without the strength of an emotion to guide them they seem to wander about aimlessly. On the other hand, the emotions manifest through the body: we know them by their effects on our heartbeat, our sweat glands, and the tension in our muscles. We recognize those of others by their posture, facial expression, tone of voice. Yet without emotions our lives are devoid of meaning, like the lamp without light. People kill, or die voluntarily, for emotional reasons; all our wars are over them. All the songs on the radio are about them. From a logical point of view, they are hopelessly immature, but there seems to be little we can do about it.

3. Time and Space

This tripartite classification of our subjective experience seems to make a certain amount of sense, but, just as with the lamp, the intermingling of categories makes understanding elusive. In dealing with the outer world we have developed precise mathematical methods that define three orthogonal dimensions of space and one of time, and specify how to locate points (events) within this continuum and how to relate the points to each other. From the basic elements of space, time, and mass, more complex concepts are constructed: velocity, force, energy. Add some atomic building blocks and you have us. Order emerges from confusion, and even the hopelessly confusing inner world can be subjugated: our fears are explained by biochemical imbalances, our thoughts are electrical patterns. The inner world has disappeared! But of course it has not disappeared; it remains the only world each of us knows. And it has intruded into physics, in that the act of observation affects what is observed; nowadays, physicists seem to be as interested in consciousness as neuroscientists. So, if in dealing with the age-old dichotomy between the inner and outer worlds, the inner world must be given equal status, the problem remains: is it possible to be more precise in defining its dimensions and measurements?

Although the outer and inner worlds are clearly related, one might even say superimposed, they are also incommensurate. Objects in the outer world can be located in the three dimensions of space, but thoughts and emotions cannot. Where then are they located? One idea is that they are located not in space, but in time. Neuroscientist Peter Fox: "The brain exists in space. But now the mind—the mind operates in time alone."[4] Gurdjieff calls time the "Ideally-Unique-Subjective-Phenomenon."[5] It has been argued that time does not exist in the absence of subjective experience. Time is required because things change, and although change is a property of the external world, the perception of it must be internal, because externally only the present configuration of objects exists, and the perception of time requires comparison with the past configuration, which exists only in our minds. Even a series of records or photographs derive their relationship as members of a temporal sequence only by the operation of our minds.

If time is the space of the inner world, then it must have more than one dimension to accommodate the full range of subjective experience. Along with the perception of change, a unique capacity of mind is the perception of possibility. Our thoughts constantly roam over different possibilities, both in the past and the future; thinking could be called the selection of some configurations of possibility from the many that constantly bubble up before our inner vision. Change and possibility are two different dimensions. Possibility refers to different possible changes. One can conceive that change could exist without alternative possibilities (the clockwork universe, predetermined to follow an immutable sequential development, was a viable scientific notion until the modern era), and one can conceive of different possibilities but no change. In either case, we would not need brains. Our brains only make sense in a world in which both change and possibility exist; conversely, perhaps change and possibility can only exist because of our—or other—brains.

Ouspensky postulated three dimensions of time: linear time, as we usually think of it, moving steadily from past to future at "one second per second"; the "line of eternity" or the perpetual existence of each moment; and the "line of the actualization of other possibilities which were contained in the preceding moment but were not actualized in 'time.'" [6] We have been discussing the first and the third of these dimensions. The second, "eternity," is more subtle. In a sense it is a necessary backdrop for the other two, for without the perpetual existence of each moment, in some form, neither change nor possibility would have any meaning.

Put another way, the measurement of time requires a standard for comparison, which is necessarily cyclical. [7,8] So there is a connection between the idea of eternity and cyclical phenomena. The projection in linear time of a cyclical phenomenon is a wave: the familiar sine wave can be constructed by projecting a point moving uniformly around a circle onto a moving plane, perpendicular to the plane of the circle. (Figure 1)

By comparison, the elaboration of possibilities resembles a tree. The tree is a ubiquitous shape in both concrete and abstract phenomena, suggesting that it has a fundamental place in the description of the world. There are plant-trees, bronchial trees, arterial trees, family trees, decision trees. Many of the neurons of our brains resemble trees. The transformations of elementary particles look like trees. Road systems and rivers are tree-like.

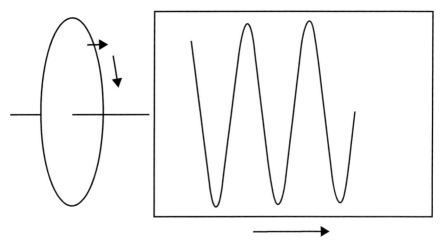

FIGURE 1

A tree represents an elaboration of possibilities in time. A family tree is a clear example. A decision tree is more abstract. A plant-tree, or a bronchial tree, existing all at once in space, is also the imprint of a branching growth pattern in time, and, for a particle of sap or air, it is also a sequence of "choices" in time, just like a road system is for us as we travel along it. But a tree is also a way to distribute a substance, such as air in the bronchial tree, over a large area, so that it can undergo a transformation by interacting with another substance, in this case blood, which has been distributed to the same interface by the pulmonary arterial tree. A plant-tree distributes chlorophyll and other metabolic machinery over the large area of its leaves so that sunlight can effect the synthesis of oxygen. A tree is a prime example of a fractal growth pattern, a way of generating a large surface area in a small volume. Our bodies are packed with trees—vascular trees, neural trees—every cell needs to be close to the vascular system and the nervous system, and that is how it is done. An organ—a lung, a kidney, a liver—is in effect a set of intertwined trees, in which substances distributed over a large interface interact with other substances, and the transformed results are re-collected into main pipelines for further distribution elsewhere in the body. So what is, on one scale, a pattern of possible choices, is on another a system for the transformation of substances.

In another one of his meaning-laden cryptic sentences, Gurdjieff said: "Time is breath."[9] Breathing proceeds in ordinary linear time, is cyclical, and involves the movement of air and blood within trees. The earth, with its covering of trees, also breathes cyclically: during the day trees produce oxygen, during the night they do not. A life can also be like a tree: as we grow up, more and more possibilities open up, and we may take advantage of a number of them—learn this and that, acquire some skills, experience many things. Then, in and past the prime of life, a life can recollect itself, pull together its achievements, distil its wisdom. Perhaps then the kernel of essence it was born with has been transformed and vivified, like blood in the lungs, for its further journey, which, according to some, consists of another cycle of life.

Mathematically, a wave, or vibration, is a circular motion projected onto linear time. But a real wave, for instance a water wave, expands in all directions. Portions of it meet other waves, obstacles, shallows. It gets absorbed, reflected, refracted; new waves are created by the interactions. A tree-like set of possibilities unfolds.

So we have a new triad: line, wave, tree, closely related to the triad we began with, of matter, energy—or life—and information. Matter clearly exists in linear time. Energy, in many of its forms, is vibration: sound, light, heat, are all waves. The concept of energy in physics also includes potential energy, such as that of an object perched in a gravitational field, and the kinetic energy of motion, as when that object falls. But let us provisionally entertain the idea that energy, at least the energy that gives things life, is closely allied with vibration. Information is related to the dimension of possibility: a piece of information is such only by virtue of being one possibility among many.

Phenomena occur on many different levels, or scales. From atoms to galaxies, on each scale, a relatively self-contained world of things and interactions exists. The division of science into relatively self-contained disciplines reflects this. On the scale of the solar system, the details of my life are unimportant, though humanity as a whole may play a role. For me, the activities of any given red blood cell are not within the domain of my perception, but the bulk properties of my blood are very important. But each level also depends on and interacts with the others, in both directions. Without cells, molecules, atoms, I could not exist, and the macroscopic structure of my body in turn constrains and directs the behavior of its constituent parts. Often there is a change in the type of law that operates on two different scales, for the same phenomenon. A container of gas obeys deterministic laws regarding temperature, pressure, and volume, but the gas molecules are moving in an essentially random way, bouncing off each other and the walls of the container, and this randomness is essential in producing the bulk properties of the gas.

On every scale time moves along, cycles repeat, possibilities unfold. There are cycles within cycles, and trees within trees. But in going from one scale to another there is typically a change in which of these aspects is important. The electrons around an atom can be considered as the vibrations of standing waves (like the vibrations of a guitar string), but for us this manifests as solid matter persisting in time. A rapidly turning propeller looks like a solid disc, and behaves, in part, like one. Linear time is measured by cyclical phenomena: the vibrations of a quartz crystal, swings of a pendulum, cycles of day and night and the seasons. The choices of a lifetime become part of an insurance company's statistics, subject to cyclical variations. The cycles of intermediary metabolism become my body's bulk supply of energy, the cycles of breathing its oxygen supply. The totality of choices made by my cells as they multiply and differentiate becomes my body itself.

Another change that occurs as one moves between levels is from analog to digital, or continuous to discrete. For the gas molecule, hitting the wall of the container is a discrete event, but for the gas as a whole, there is a smoothly varying pressure. A wave, or its equivalent, the circle, is the most continuous of phenomena. But given a string tethered at each end, only those discrete frequencies can sound for which an integer number of half-wavelengths fits into the confines of the string. (Figure 2) This produces a discrete series of harmonics. The same considerations apply to tubes of fixed length, and every pennywhistle player knows that when she blows a little harder, the change from one harmonic to another is abrupt, without the frequencies between the two notes being sounded.

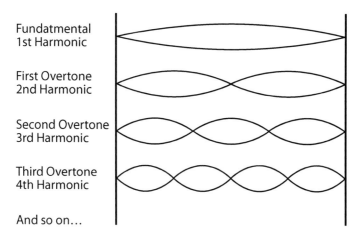

Fundatmental
1st Harmonic

First Overtone
2nd Harmonic

Second Overtone
3rd Harmonic

Third Overtone
4th Harmonic

And so on…

FIGURE 2

While the wave is a prototypical continuous phenomenon, a sequence of choices, or a tree, are distinctly digital. Information is also digital in nature: it is conveyed by sequences of discrete sounds, or marks on paper. The mark derives its information value by being a choice among all the other possible marks. As long as it can be uniquely recognized, its precise shape does not matter, as in different typefaces or handwritings. And the code is an arbitrary one, as testified to by the profusion of languages. Interestingly, music, which is essentially wave-like, continuous, and speaks to the emotions, is much more universal across cultures, and a harmony which is more than slightly off key, though still recognizable, does not have the same effect as one which is in tune.

We have elaborated on aspects of what may represent three dimensions of time, but may remain unconvinced that three dimensions of time are in fact a necessary postulate. After all, science has done very well with one time dimension, even managing to deal with waves and possibilities. But has it? For one thing, science is entirely silent on the inner world; there is no place for it within the framework of science, only for its "outer" correlates, such as the brain, hormones, etc. Further, modern science has come upon some difficult paradoxes in relativity and quantum theory, which may point to a need for more dimensions.[10] The current mathematical structure of physics apparently does not allow more than one dimension of time,[11] but at least one prominent physicist, Steven Hawking, has actively considered a second time dimension.[12]

Just as nothing exists in space that is not three-dimensional, although we can *conceive* of two- or one-dimensional objects, nothing can exist in the world of our inner experience without the three dimensions of time. The existence of different possibilities would in a sense not be real if there were no possibility of change (although we can conceive of it, just as we can conceive of a dimensionless point or a one-dimensional line), and change would also not be real if there were not different possibilities; neither could exist without the eternal existence of everything. And, just as a translation or rotation of spatial axes causes an intermingling of the three spatial dimensions, a change in perspective changes the projection of experience onto the three time axes: what was possible yesterday may be actual today, what was actual yesterday is a memory today. Yet the parallels are not complete; we seem more restricted in our movements in time than in space. Possibilities for tomorrow exist not just in our minds, but also in potential actualities, whereas possibilities for yesterday exist only in our minds, it seems. Also, space is isotropic: the three directions of space are equivalent; there is nothing to distinguish one from another, and the specific coordinate axes that we choose for our measurements are arbitrary. The three dimensions of time as postulated above, on the other hand, seem quite different one from another. However, it could be argued that in actuality, in any specific location, space is not isotropic. On the surface of the earth, for us and for all things with mass, the vertical direction is qualitatively very different from the two horizontal ones; our up and down movements are quite constrained, by gravity, compared to north-south or east-west movements. Even north-south is different from east-west in terms of our relationship to the sun, the earth's direction of rotation, its magnetic field, etc. And in fact anywhere in the universe one direction is different from another in this way, because of the existence of centers of mass, celestial bodies, all of which both rotate and revolve. This is not trivial—all phenomena are affected by these anisotropies. The concept of isotropic space is an abstraction, which serves us well for making calculations. There is a hierarchy in this respect: light, for instance, having no rest mass, is not restricted to the surface of the earth in the same way as we are; it belongs to another level of things. Yet it—electromagnetic energy—is what animates us, so we also partly belong to another level of things. And light, which has a different relationship to our space, also has a different relationship to time, as shown by Einstein.

4. The Three Brains and Time

One way to look at our three brains is in terms of their relationship to these three dimensions of time. Time as we ordinarily think of it, flowing steadily from the past into the future, the world taking a unique configuration at each moment, has proved sufficient for describing the outer world, at least in most of its practical aspects. Our bodies, at the interface between the outer world and our inner experience, live in this dimension of time. Over the long term, after an initial period of growth, decay inevitably sets in: the wrinkles get deeper, the muscles weaker, the bones more brittle, the mind less nimble, until homeostasis can no longer maintain us against the forces of dispersion and the body dies, its elements scattering in the wind. Entropy inevitably increases. In the shorter term, our bodies are subject to the laws of material things: they can only be in one place at one time, movements must be performed in a sequential fashion, and sensations appear sequentially in accordance with the stimuli that excite them. The "moving-instinctive" brain operates primarily in this mode, utilizing sequential chains of reflexes and preprogrammed patterns of movement.

The mind knows no such limitations. It roams freely in the past and the future, and among many possibilities at each juncture. I cannot undo an action, but I can without any difficulty imagine not having done it. I can remember the past, and imagine what might have been if I had done something different. The mind is generally much more occupied with these elaborations than it is with "what is actually happening now." Most importantly, there is no mental barrier between the future and the past, or between what has happened and what might have happened. No alarm signals that thought is passing through one of these boundaries. The part of our minds we live in most of the time has no limitations in these directions. And this is also its strength: imagination makes us capable of building bridges, playing chess, and planning for the future. This mode of functioning is reminiscent of Ouspensky's third dimension of time: "the line of the actualization of all possibilities."

Emotions seem to live in "eternity." Certain aspects of emotional life make sense if we regard emotions as especially related to this dimension of time. For our emotions always have the subjective quality of being "forever." Being in love is forever. Even my moods have this quality. In a bad mood I cannot conceive emotionally of any other. Suicide in the face of depression is testimony to this. I have to *think* to see that another mood is possible, that in fact I was in a different mood before and inevitably will be again, because thought is oriented differently in time. This difference is striking when I try to reason with a friend about an emotional problem of his; I am thinking, and he is feeling, two completely different ways of looking at the same situation.

We must always keep in mind that we are postulating three dimensions, not three separate entities in the usual sense, either material or psychological. Gurdjieff says, in speaking about the three brains:

> We must understand that every normal psychic function is a means or an instrument of knowledge. With the help of the mind we see one aspect of things and events, with the help of emotions another aspect, with the help of sensations a third aspect. The most complete knowledge of a given subject possible for us can only be obtained if we examine it simultaneously with

our mind, feelings, and sensations. In ordinary conditions man sees the world through a crooked, uneven window.[13]

And, in one schematic categorization, Gurdjieff subdivides each of the three brains again into three, speaking for example of the "intellectual part of the moving center." Any real object in space belongs to all three dimensions, and any inner experience belongs to all three dimensions of time, though it may lie predominantly along one of them. The moving center, in catching a ball, does not just use a sequential linear pattern of movements, but also a repetitive comparison of possibilities (repeatedly comparing where the ball and hand are relative to each other, to where they should be, so that they will meet) that is like the mode of functioning of the intellectual center. And our usual thoughts move automatically from one to another in a fashion that may be closer to the mode of operation of the moving center than to the true potential of thought. That our thoughts, emotions, sensations, and actions are interrelated is obvious; they are so interrelated that their status as three separate aspects of our experience can be difficult to perceive. Part of the difficulty may be that we are dealing with dimensions, not things. Another part may be because we do not usually live (inwardly) simultaneously in all three time dimensions.

So far I have mostly relied on abstract arguments and on the analysis of inner experience in arguing for the existence of three brains, each related primarily to a postulated dimension of time. To what extent do these concepts fit with the scientific information available about our nervous systems?

5. A Quick Lesson in Neuroscience

The most important difference between a living animal and a dead body is that the electrical system has been turned off; "the lights have gone out." Of course if blood doesn't flow—frequently the cause of death—the electrical system will not work; electricity needs to be generated, and the generator needs fuel. Every cell in the body has an electrical potential across its membrane and cannot function without it. Broadly speaking, all our metabolic energy is ultimately electrical energy.

In the nervous system, electrical energy is used not just to move molecules about, contract muscles, pump air or blood, filter wastes, or provide for general maintenance, but to process and exchange signals on a massive scale. Signals to inform us about the environment, coordinate activities in different parts of the body, and to think with, somehow. How these signals provide for thought, perception, and consciousness remains the central mystery of neuroscience, despite a wealth of information about the nervous system.

Most nerve cells—neurons—carry signals, which consist of a change in the electrical potential across the cell membrane, called the action potential. These signals travel along the extended processes of the nerve cells, called axons (from the spinal cord to the foot is perhaps the longest), at up to 100 meters per second. The information carried by the axons is encoded in the frequency and temporal pattern of the action potentials. Generally at one end of the neuron, the cell body and dendrites, a large number of connections, called synapses, are made from other neurons, and at the other end the axon connects in turn to one or more other neurons (or directly to other kinds of cells, such as muscle cells). (Figure 3) The connections are of many kinds but typically involve a chemical messenger, released from the end of the axon and influencing the electrical properties of the recipient cell. One can picture the cell body as a place, a smoke-filled room if you like, where the incoming information is weighed, according to strict electrical criteria of course, and

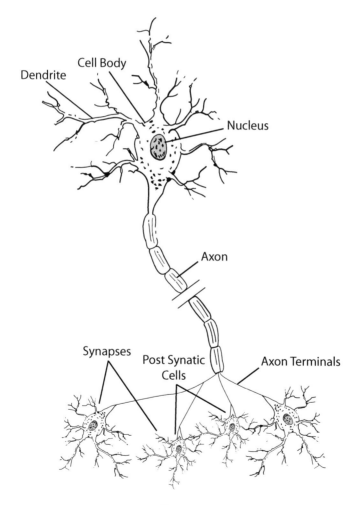

FIGURE 3

decisions are made as to what kind of message to send out along the axon, namely what frequency, and pattern, of action potentials.

The nervous system can be divided anatomically into many distinct parts. The most obvious division is between the peripheral nervous system, lying outside the confines of the skull and spinal column, and the central nervous system, lying within these bony structures. The peripheral nervous system connects the central nervous system with the rest of the body, and consists of input nerves (bundles of axons) that convey information from the sense organs into the central nervous system, and output nerves that allow the central nervous system to control muscles and glands. The peripheral system is further subdivided into the voluntary and autonomic systems, the former having to do with muscles under voluntary control—the muscles of movement of the body—and the latter with the functions described above as instinctive—digestion, blood flow, glandular secretion. The autonomic part of the peripheral nervous system, unlike the voluntary part, makes complex interconnections outside the central nervous system, in comglomerates of neurons called ganglia, and so is not merely a relay system.

To look for what can be called a brain, or brains, we usually go to the central nervous system. (Figure 4) The term "brain" is conventionally singular, and applies to the contents of the skull, but what defines a brain is not often specified. A brain can be defined in various ways, all of which are satisfactory, at least as a first approximation. Functionally, and in modern terms, it can be thought of as a structure that carries out complex information processing tasks. Physiologically, the substrate for this kind of function seems to be a network of large numbers of massively interconnected neurons. Gurdjieff describes a brain as an instrument of knowledge. "Complex information processing" could be regarded as a necessary, though perhaps not sufficient, condition for knowledge. The brain is also the coordinator of behavior, or manifestation. At least this is the case in higher animals, for the behavior of a stone, or a plant, is not thought of as under the control of a brain. So we have to say: "sufficiently complex behavior," or "behavior which demonstrates evidence of underlying knowledge." At the edges of these definitions are all of the difficulties that have confronted philosophers and scientists for centuries, and continue to baffle us: what is knowledge, consciousness, action? But for now let us go back to the central nervous system, where most of these elaborate networks of nerve cells are found. The intricacy is staggering, with hundreds of conglomerations of neurons, and hundreds of axon bundles connecting these conglomerations with each other and with the peripheral nervous system. It is said that any two neurons, as far apart as you like both in space and in function, are likely to be connected to each other through no more than five intervening neurons (five "degrees of separation"). But in this confusion subsystems can be discerned.

Many categorizations are possible, but a broad subdivision into three parts, as put forth by MacLean, does not seem arbitrary.

The spinal cord and part of its extension in the back of the head, the "brainstem," together thought of by MacLean as the "reptilian brain," correspond to Gurdjieff's moving-instinctive center. Almost all the nerves of the peripheral nervous system feed into to this part of the central nervous system, and large portions of this structure are occupied by bundles of axons that connect the peripheral nerves to higher levels of the central nervous system. However, it is not just a relay station. A wide variety of elaborate reflexes are organized at this level, and it can be considered a brain on its own. The nervous system operates in a hierarchical fashion, like any good organization. Complicated patterns of movement, for instance, are organized at the spinal level, and are set into motion, modulated, and refined by input from higher levels. The capacities of the spinal cord are most easily appreciated in lower animals which are less burdened by higher organizational levels—a chicken without its head can run—but the same pattern exists in humans. It is interesting to consider that the withdrawal of my hand from a hot stove, mediated by a spinal reflex, occurs before I am consciously aware of the

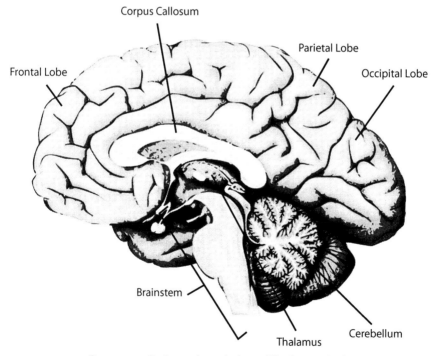

FIGURE 4 • Brain cut through the middle, front to back.
Midline aspect of cerebral cortex, rest of structures cut in half.

heat, even though I say without equivocation "I took my hand off the stove." Likewise, instinctive functions are largely organized at this level, particularly in the brainstem, and in the case of injuries which destroy the higher portions of the brain but leave the brainstem intact the body can continue to breathe, maintain blood pressure, and remain alive. However, consciousness, perception, and thought are, to all appearances, lost.

At the other end of the nervous system is the cerebral cortex, the highly convoluted surface of the brain, most developed in humans. It consists of a sheet of neural tissue several millimeters thick, the "gray matter," composed of nerve cells with elaborate local interconnections, many of which are also connected to other parts of the nervous system through longer axons which make up the "white matter," under the cortex. Connections are made through the white matter to adjacent parts of the cortex, to distant parts of the cortex, to the cortex on the opposite side, to subcortical conglomerates (nuclei) of neurons, to the cerebellum (described below), and to the brainstem and spinal cord.

The activity of the cerebral cortex seems to be the substrate for the *contents* of ordinary consciousness, although the "energizing" of consciousness is mediated by deeper structures within the brain, which in the aggregate have been called the reticular activating system. These structures control the *state* of consciousness (waking, various kinds of sleep), and if sufficiently disrupted result in coma. Discrete portions of the cerebral cortex are devoted to visual, auditory, and sensory perception (a number of areas to each), to multisensory functions such as spatial orientation, to language (different areas are specialized for language output, language comprehension, reading, and writing), to control of movement, to planning of movement, and to other "higher functions." The evidence for this is abundant: Damage to various parts of the cortex produces deficits in specific functions. Electrical stimulation (during neurosurgical procedures or seizures) produces sensations, movements, complex hallucinations, memories, or thoughts, depending on the area stimulated. Specific mental tasks can be correlated with increased metabolism in specific areas, as measured by special scans. On the other hand, specific functions are not rigidly compartmentalized. The brain seems always to be active as a whole, but certain areas and groups of areas are clearly specialized for carrying out certain activities.

The cerebral cortex is the "neomammalian" brain of MacLean, and seems to correspond, at least in part, to the thinking center of Gurdjieff.

The parts of the central nervous system involved in emotional life tend to be situated in a sort of middle layer. They include portions of cortex that are simpler than the six-layered neocortex of the "thinking center," collections of neurons not arranged in a cortical structure but grouped into nuclei, and portions of the upper brainstem. The structures forming the "reticular activating system," which regulate the state of consciousness, can be considered to be part of this middle layer. This regulation is possible because of widespread connections to the entire cortex, whereby small collections of neurons can modulate the activity of large numbers of cortical neurons. The *state* of consciousness is reflected in the bulk electrical activity of the brain, as recorded from the surface (electroencephalogram, EEG), whereas the *contents* of consciousness are only subtly reflected in this activity. The EEG recording shows rhythmic patterns, and the prevailing rhythms change in different states, for instance, in sleep and coma. (Figure 5) While much less is known about states of consciousness beyond the waking state, recordings of meditating yogis and Zen monks tend to show differences in the EEG compared to the ordinary waking state. Some of the structures of the middle layer are also critical for conscious memory; damage to relatively small structures can completely disrupt the laying down of experiential memories, although old memories are retained, having been thoroughly imprinted on the cortex. These three aspects of inner activity—

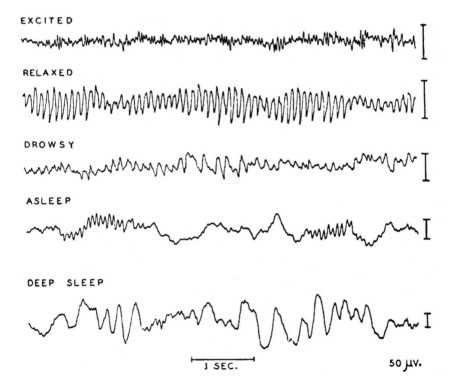

FIGURE 5 • EEG in different states of alertness

level of consciousness, emotion, and conscious memory—are also related to each other in our experience. In sleep we rarely form memories of what is going on. Emotional intensity tends to make memories vivid and detailed, like relivable experiences. States of consciousness beyond the waking state are accompanied by, indeed characterized by, feelings which we usually lack: awe, wonder, a sense of mystery, of being alive, of being connected with everything, of being present.

6. Brain and Function

Beyond the basic understanding of how neurons function, much of the information accumulated about the nervous system is fundamentally anatomical: which neurons are connected to which others. For instance: our eyes move. There are six muscles attached to each eyeball. These muscles are supplied by nerves, coming from the brainstem. The neurons making up these nerves are connected in turn to neurons in other parts of the brainstem that coordinate eye movements. These areas are further connected to many other areas, including those that have to do with vision. This makes sense because we move our eyes in response to visual stimuli. The eye movement control system is also connected to the balance organs in the inner ear, for we have reflexes that make our eyes move opposite to movements of the head, to help stabilize the visual image in the face of constant body movement; a person without these reflexes experiences constant bouncing of visual images as they walk. Much of our knowledge of the nervous system is of this sort. The details are fascinating and important, but in a way it is all obvious. If something rolls, it must have some kind of wheels. That we have muscles attached to our eyes and not strings, that the nervous system transmits information via traveling electrical potentials and not microscopic carrier pigeons, that neurons are massively interconnected with chemical transmitters mediating many of the connections, these are the fundamental questions that have been answered. But how we make sense of what we see, how we plan a series of actions and then carry them out, and, above all, how we have the capacity for consciousness, these are as yet unanswered questions. The nervous system is intimately connected with the mind, with the inner life we all experience, but the nature of the connection remains almost as mysterious as it was three hundred years ago when Descartes tackled it. Some organizational principles are beginning to be elucidated, however, and for the purposes of this particular essay, we could ask: if we are postulating three "brains," each of which relates to the world in a particular way, what is known about the general mode of functioning of each of them? How are they similar, and how different?

7. The Moving Brain

One could say that the motor system functions as a hierarchical series of neural loops. At the lowest level are simple reflexes. The simplest of these is the stretch reflex, such as the extension of your leg elicited by your doctor when he taps you below the knee with his rubber hammer. The sudden lengthening of the thigh muscle produced by the hammer blow elicits a reflex contraction of the muscle. This reflex involves only two sets of neurons, sensory ones (from receptors in the muscle that sense stretch) going into the spinal cord, and motor neurons from the spinal cord out to the muscle; the sensory neurons excite the motor ones. One can see that a reflex like this is useful in standing: gravity stretching the thigh muscle and making the knee bend results in contraction of the same muscle, keeping the leg straight. A similar reflex in the biceps results in extra contraction when a load is placed in the hand. The withdrawal reflex, as from a hot stove, is another example of a useful thing to have. This reflex is also mediated by the spinal cord, though it is a bit more complicated than the stretch reflex. But then suppose I step on a hot coal. Withdrawal reflexes make my leg withdraw and stretch reflexes make it plant itself more firmly on the ground. Some higher-level decision making is clearly needed. One could suppose that it's more important not to get burnt, so the withdrawal reflex should be made stronger. But obviously in real situations a variety of responses could be appropriate. If withdrawing my leg would mean falling into the fire I should not withdraw it; if I do withdraw it, a whole symphony of other reflexes allow me to stand on one foot. There are multiple hierarchical levels of regulation, with simple reflexes as one of the basic building blocks.

Aside from reflex circuits, the spinal cord contains pattern generators. Coordinated four-limb walking movements in an animal can be generated by the spinal cord, disconnected from the rest of the central nervous system, under certain circumstances. These pattern generators do not simply consist of chains of reflexes, for they do not require sensory input or feedback. However, they are modified by sensory feedback, and obviously also highly regulated by other levels of the nervous system.

Perhaps the highest level of this hierarchical system of neural loops is in the cerebellum. (Figure 4) The cerebellum looks like a small brain on its own, and sits astride the brainstem in the back of the head. Like the larger forebrain, it has a cortex and underlying white matter, and deeper nuclear aggregates of neurons, though it differs from the forebrain in lacking the same degree and complexity of interconnections between different cortical areas. This "little brain" coordinates all the movements of the body, making them smooth, efficient, and integrated. Among its many inputs are those from the spinal cord, some of which convey direct sensory information, while others reflect the activity of spinal circuits. It also receives a huge input from areas of the cerebral cortex concerned with movement, planning of movement, and sensation. One of its functions seems to be to compare intended movement with the actual result, and to bring the two into harmony. At this level, there is also a motor memory. The incredible coordination necessary for the complex skills animals display requires both responsiveness to the needs of the moment, and efficient habits. Skills are refined until they become effortless and automatic. A patient with a diseased right half of the cerebellum (which coordinates primarily the right limbs) said: "The movements of my left arm are done subconsciously, but I have to think about each movement of my right arm."[14] Adjustments to sensory mismatches are also mediated by it, as for instance when a radically different pair of glasses at first causes dizziness and disorientation and after a few weeks everything seems normal again. However, patients with damage to the cerebellum do not appear to have deficits in consciousness, perception, language, or thinking.

The cerebellar cortex has a structure so regular it has been called almost crystalline. (Figure 6) The operations it performs, for a wide variety of movements of every kind (and possibly also on plans for movement before they are actualized), seem to be essentially the same, despite differences in the movements themselves. For this reason it has been called a "neuronal machine," and great effort has been expended on finding out what kind of

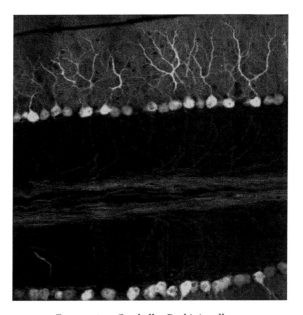

FIGURE 6 • Cerebellar Purkinje cells

machine it is. Much is known about the basic neural elements and how they are connected within the cerebellum. The type of deficit occurring when the cerebellum is damaged is also quite uniform; for instance, the incoordination of eye movements resembles that of limb movements. Precisely how the machine does its work is still unsolved, but its uniformity is interesting. If we knew nothing about the inner workings of such functionally different machines as televisions, amplifiers, calculators, and electric pianos, the discovery that they all contained similar circuits of transistors, resistors, and capacitors would take us a long way toward both a practical knowledge and a deeper understanding of what kinds of machines they are.

The conceptual framework that has been most successful for understanding the motor systems is that of control systems engineering, the design of complex machines and assemblages of machines. Hierarchical arrangements of feedback loops, comparison of intended output and actual output, transformation of data into rates of change, and integration of input over time—these are the kinds of terms used. In fact, some researchers in such fields as the control of eye movements have engineering backgrounds. Gurdjieff's characterization of man, in his usual state, as a machine, is not only correct at this level of our functioning, but this is the appropriate mode at this level. The habitual and automatic, non-conscious functioning of our bodies as they walk and talk and perform all sorts of incredibly skilled actions is highly desirable; if we had to think about each movement of our right arms, like the patient, we wouldn't be able to do much.

8. The Thinking Brain

Of course, the control of movement does not end in the cerebellum; patients with strokes affecting motor areas of the cerebral cortex (Figure 4) are effectively paralyzed, although certain reflexes can still be demonstrated. In view of the amount of circuitry devoted to movement below the level of the cerebral cortex, the degree of impairment from cortical damage is striking. As mentioned above, lower animals are less dependent on their cerebra. At the level of the cerebral cortex movement blends with thought. Areas that specialize in planning movement and coordinating sequential movement strategies (premotor areas) are found anterior to those that are more exclusively concerned with muscle movements alone (motor cortex). A dramatic illustration of this is found in experiments using scans that show which areas of the brain are most active under different conditions. In the performance of a simple automatic movement, the motor cortex is especially active, whereas with a more complex sequence of movements both the motor cortex and premotor areas are active. Mental rehearsal of the task without actual movement is associated primarily with activity in the premotor areas.

Thus as we move forward in the frontal lobe from the primary motor region of the cortex we encounter areas that seem to be concerned with progressively more abstract aspects of motor control. With sensation also, parts of the cortex that receive direct sensory input are next to others that carry out more sophisticated analyses, and these in turn abut regions that deal with multisensory integration—aspects of perception. The higher level sensory areas are connected to the higher level motor areas. Where, along these routes, would our postulated moving center end and our thinking center begin? There seems to be no clear division. A related question is: Lower mammals, which we have speculated are endowed with only two "brains," and which give every evidence of being a quantum leap away from us in conceptual capacity and probably also in capacity for consciousness, have cerebral cortices, although in dramatically lesser amounts and with much less devoted to higher level associative functions. What then constitutes the third "brain"? Evolutionary theory denies that there is a quantum leap between animals and man, although common sense suggests one. A number of quantitative differences other than the amount of cerebral cortex exist between monkey and human brains. There is also at least one difference that could be considered qualitative: The appearance of the brain, like that of the external body, is more or less mirror symmetric, and the cerebrum consists of two separate "hemispheres," connected by a massive tract of nerve fibers called the corpus callosum. In man, to an extent not seen in other mammals, the right and left cerebral cortices are differentially specialized with regard to higher level functions. We will return to this subject after discussing some other aspects of cerebral cortical function.

9. Substance, Energy, and Information

Although to a lesser degree than the cerebellar cortex, the cerebral cortex is also striking in its uniformity. This is all the more surprising here. The coordination of movement performed by the cerebellum is similar for the various kinds of movements that are coordinated, even though they differ in the body parts involved and in the size and type of the movements. But in the cerebral cortex different areas are devoted to functions that seem to differ markedly—language, fine finger control, and recognition of faces, to name three almost at random. Despite this, the cellular arrangements are quite similar in different parts of the cortex, although differences do exist in the prominence of certain layers or certain cells in different parts, and these differences correlate to some degree with different functional subdivisions. What the cortex does with regard to transformations of electrical signals is not broadly understood (although many bits of information have been amassed), but the similar arrangement of cells throughout the cortex suggests that it—like the cerebellar cortex—can be regarded as an organ, like a lung or a kidney, which effects a transformation on a substrate. The lung, as mentioned above, takes in venous blood and air through large pipelines, spreads them over a large surface where they interact across a membrane so that the blood is oxygenated, whereupon it is collected back into other pipelines for distribution to the body. The cortex can be regarded in a similar way, taking in neural information via large fibers tracts, transforming it in a specific way, and then sending it out along other tracts for further processing.

(There is a marked difference between the cerebral and cerebellar cortices, alluded to above, in that the cerebral cortex has many more intercortical connections. This difference must reflect a major difference in the two cortices' modes of functioning.)

However, a major difference between blood and air on the one hand, and "nervous energy" on the other, is that the former have a "bulk" quality about them: a particular small portion of blood in a large blood vessel, say near the edge of the vessel, could be interchanged with another, say in the middle, without altering the nature of the phenomenon. Correspondingly, blood vessels and bronchial tubes are pipelines without longitudinal subdivisions. The information content of a nerve fiber tract, on the other hand, depends on the patterns of impulses in the individual nerve fibers that make up the tract. The optic nerve, for instance, which leaves the eye and carries information to the next relay in the visual system, contains 1.2 million individual axons in a bundle a few millimeters in diameter. Different axons come from different portions of the retina (the light-sensitive sheet in the back of the eye), and the spatial pattern of the image has to be kept in order, or the most important part of the visual information would be lost. So one could not interchange nerve fibers from different portions of the nerve without fundamentally altering its "contents."

In general, "energy" and "information" have this same qualitative difference between them, energy having a "bulk" quality and information a "specific" quality (and, as discussed above, energy having an analog and information a digital character). Mathematically, however, information can be converted into energy by thermodynamic relationships that relate energy with order and disorder. And, of course, blood and air are highly ordered substances microscopically. It is by virtue of very specific atomic and molecular structures that energy is "carried" by them; biochemical transformations consist of changes from one structure to another, which either release or absorb energy. Metabolism is a large set of interrelated changes in molecular structure, arranged in complex networks of interacting sequences.

So the order evident in the pattern of activity in nerve fiber bundles is order of a different kind and on a different scale. The difference in kind is related to the difference in scale. The order of a molecule can be regarded (to some extent) as a spatio-temporal arrangement of subatomic particles, but to our perception a molecule is less than a dimensionless point. We are aware of molecules only in large quantities and only through their aggregate manifestations: heat, color, wind, food. For our perceptions, on the other hand, the spatio-temporal arrangement of the components—the pattern of impulses in the axons of a nerve fiber bundle—is the most important feature.

From the point of view of an individual craftsman in a furniture factory, who makes table legs on a lathe, the details of each leg are the important units for consideration; to the foreman of the shop, individual table legs are subordinate to the completed tables. To the manager of the factory, tables are just units of the bulk flow of goods produced by the factory, which flow is mathematically convertible into money. At the level of the national economy, the money generated by the factory becomes a unit of information.

Even in perception, different aspects operate on different scales. Gross movement and shape in the periphery of the field of vision provide for general orientation. Fine analysis of shapes by central vision serves recognition and manipulation of objects. Light and dark cycles have effects on endocrine secretion and emotional tone, as in winter depression and "spring fever."

From the point of view of the solar system, if it is conscious, human perception might be regarded as a transformation of one kind of energy into another.

10. The Thinking Brain, Part 2: The Transformation of Sensory Information

The beginnings of the process of transformation of sensory information have been elucidated for vision. The retina (Figure 7), which is an outpost of the brain, and like the cerebral cortex has a multilayered neuronal structure, takes the visual image projected onto it by the lens system of the eye, and transforms it to correct for the huge variations in illumination that exist in the world. This is like adjusting the sensitivity of photographic film, but can vary from one part of the retina to another. The retina also enhances contrast at the borders of visual objects and de-emphasizes gradual and mild variations of illumination within relatively uniform areas. These transformations are obviously useful as a first step in breaking down the visual scene into objects. The transformed scene is then carried by the optic nerves of the two eyes to the lateral geniculate body, a subcortical nucleus of cells in the thalamus, whose transforming function is largely mysterious but may be related to visual attention. Then the visual information is relayed to the occipital cortex in the back of the brain. Before they reach the lateral geniculate body, the two optic nerves mix in such a way as to bring superimposed views from the two eyes into register, and ultimately the right half of the visual scene (from the two eyes) is brought to the left occipital cortex and the left half to the right occipital cortex. (Figure 4)

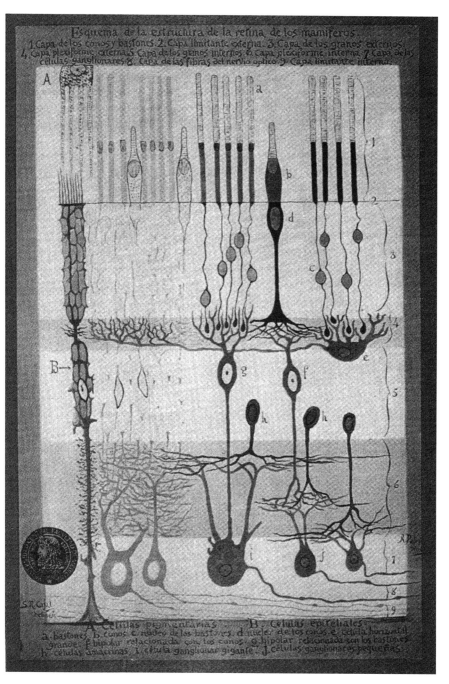

FIGURE 7 • The Retina (from Ramon y Cajal)

The occipital cortex carries out other sets of transformations. One of them is to compare the signals from the two eyes in such a way that depth is perceived, depth perception being dependent on the slightly different views that the two eyes have of a three dimensional object.[15] Another is to break down the visual scene into lines, or edges, of various orientations. Another is to break it down into components of different spatial frequencies.

Spatial frequency refers to the spacing of regular arrays of stripes, or spots. For example, a striped shirt may have a frequency of two cycles, or stripes, per centimeter, whereas bannister railings might be spaced at ten cycles per meter. Mathematically, any spatial pattern, even an irregular one, can be analyzed into a set of superimposed spatial frequency components, a process called Fourier analysis. That the auditory system performs a Fourier analysis on sound waves has long been known, since we can hear the individual notes that compose a chord. That the visual system does something similar is not so obvious, but it seems to be the case that this occurs as one of the early aspects of visual processing. The rapid discrimination of different visual textures, for instance of a cornfield from the row of trees behind, the sky above, and the ground in front, is related to this.

In any case, transformations of visual input are performed by the occipital cortex, and the transformed information is then relayed to other places. It is known now that a number of operations of this sort occur in parallel as well as serially. The further analyses of form, of motion, and of color, are carried out in different portions of cortex at the same time. Relevant information from the occipital cortex is relayed to these areas, and, interestingly, these areas project back to the primary visual cortex. One idea about cortical function is that, by these reciprocal connections, a process of fine-tuning can occur.[16] It is as if in our furniture factory the table were assembled with the parts only crudely fashioned, then disassembled. Then the parts were sent elsewhere for further shaping, then reassembled, then sent back for sanding, then reassembled, and so on. At some stage, the results of a number of different visual analyses join with similar results related to touch, hearing, and other information, in cortical areas known as multimodality areas.

The transformations occurring in the cortex can be thought of as comparisons. Even in the retina, the intensity of light is being compared between adjacent areas. A dark area next to a light area looks even darker at the edge between the two. In the visual cortex the inputs from each eye are compared to create depth perception. In many parts of the brain, input from one hemisphere is compared with input from the other. These comparisons are made possible by a three-dimensional spatial arrangement of neuronal connections. In the retina the two-dimensional image projected onto the light-receptor cells (rods and cones), a mosaic of points of different intensity and color, is relayed to the next layer of cells, which forms as it were another two-dimensional screen just behind it (anatomically, it is actually in front) by connections which are perpendicular to the plane of each layer. (Figure 7) Horizontal connections, in the plane of the layers, effect the comparison of adjacent light and color intensities, so that the second image is a transformed version of the first. Multiple comparisons of this sort are probably the basic mode of operation of the cerebral cortical machine—its transistors, as it were. One of the effects of the inputs to the cortex from the "emotional" parts of the brain seems to be to change the strength of these comparisons, for instance making the contrast enhancement at edges stronger or weaker.

A comparison is also a decision—a branch point in a tree. The cerebral cortex is a set of such trees; it even looks like a forest of microscopic trees. The emotional brain interacts with this forest by biasing its branch points one way or another, to change the output of the trees. This set of interactions occurs in a plane that is physiologically perpendicular to the planes of the branchings; if the envelope of the tree is regarded as a cone, a set of biasing operations would make a circular cross section through the cone.

Our thinking, viewed internally, also seems to consist of multiple levels of comparison, occurring both in series and in parallel. An abstraction is at the top of a pyramid of comparisons.

12. The Split Thinking-brain

To return to the question of the uniqueness of the human brain: The two cerebral hemispheres are largely similar in structure, each related to the opposite side of the body (for movement and sensation) or the opposite side of the world (for vision, and to a lesser degree, hearing). (Interestingly, the cerebellar hemispheres project to the ipsilateral (same) side of the body, and the projections of the emotional brain to the spinal cord and thence to the autonomic ganglia are also ipsilateral.) But in humans, to a degree seen in no other mammal, the two cortical hemispheres are partly asymmetric in function, although similar in structure. Even at the motor level there is marked asymmetry, since the preferred hand (in most people the right, controlled by the left hemisphere) is generally much more skilled at making fine movements. And in the vast majority of right-handed people, and even in 75% of left-handed ones, the left hemisphere has almost exclusive domain over most aspects of language.

(The many qualifiers—most, almost, etc.— are unfortunately necessary for accuracy, for obviously there are left handed people, and in some the right hemisphere is dominant for language. These and many other exceptions are interesting in their own right, but do not, I believe, change the points I wish to make.)

Most researchers agree that the right hemisphere is typically more skilled than the left in visuo-spatial analysis. The left hemisphere also calculates, and employs logic; the right may be more capable of musical appreciation, including the appreciation of the musical aspects of speech, such as tone of voice and inflection. The fundamental difference in the mode of operation of the two hemispheres, if there is one, is the subject of much speculation, and to some extent this speculation has gone way beyond the facts, but it remains intriguing and may be important to our understanding of what makes us uniquely human. Some of the dichotomies that have been proposed (left versus right hemisphere) are: sequential-simultaneous, temporal-spatial, digital-analogical, logical and analytical-gestalt and synthetic, and rational-intuitive.[17]

The corpus callosum, the massive nerve fiber tract that connects the two hemispheres, connects for the most part near mirror-image points. It is most developed in humans, and is not fully functional until puberty. Its function is only partly understood. It clearly serves to some degree to coordinate the two hemispheres, and to transfer information from one to the other. But it may also mediate interactions between the two hemispheres that make it possible for each to function in its specialized way.

Although the fundamental mode of interaction of the hemispheres remains speculative, some examples may provide a sense of it. As mentioned above, the right hemisphere, though endowed with poor language capabilities, seems superior at understanding the musical inflections of speech, through which much information is conveyed, both complimentary to and separate from the strict meaning of the words themselves. Even more intriguingly, patients with right hemisphere damage may have difficulty in understanding both humor and metaphor. They may "have clear difficulties in integrating specific information, in drawing proper inferences and morals, and in assessing the appropriateness of various facts, situations and characterizations."[17] A deficit in the ability to form concepts has been found in such patients.

It is interesting in this regard to take note of the fact that what we call thinking takes place largely in words and pictures. It is difficult to conceive of thinking without either one. Smell, taste, and touch can also be evoked internally, but to a lesser degree, and seem to be more incidental to the process. (Although the idea of cinemas that provide touch and smell stimuli as well as visual and auditory ones has been contemplated for decades, there seems to be no great demand for them.) Similarly, a presentation typically involves slides and speech. Unimodal activities, such as listening to a story or reading, always evoke strong mental images, without which we cannot imagine the activity taking place. Thus the specialization of the left hemisphere for language and the right for visuo-spatial analysis has a parallel in common experience. The right hemisphere is not devoid of language capacities, nor the left—even less—of visuo-spatial understanding, but their respective specializations in each realm and the re-integration across the corpus callosum may reflect a fundamental property of human brain functioning that makes us capable of conceptual thought in a way that hardly exists in other animals.

Patients whose corpus callosum has been surgically cut, for medical reasons, have been extensively studied and have provided some of the information about hemispheric functions. Although such patients appear at first glance to behave quite normally, retaining both language and other mental capacities, they show significant defects in both memory and the ability to sustain attention.[18]

These musings raise difficulties for our simple three-brain scheme, or at least indicate that things are not so simple. Many of the putative specializations of the right hemisphere seem related to the emotional life: music, prosody, humor. And the integration of the two hemispheres seems related to the integration of the three brains that may make our capacity for consciousness possible.

13. The Emotional Brain.

Gurdjieff did not place the center of gravity of the moving brain in the portions of the cerebral cortex devoted to movement, even though damage to these areas results in paralysis, but in the spinal cord. Similarly, he did not center the emotional brain between the spinal cord and the cerebral cortex, though much of what we consider emotional activity seems to depend on these areas. He considered the emotional brain to be located primarily in the ganglia of the autonomic nervous system. The autonomic nervous system is the part of the peripheral nervous system concerned with what we might call instinctive activities—pupillary constriction and dilation, sweating, blood flow control, heart rate, intestinal activity, erection, and ejaculation, to name a few—which are largely not under voluntary control. Like other parts of the peripheral nervous system, the autonomic system connects to the spinal cord and brainstem, where it comes under the influence of higher portions of the nervous system. However, unlike the sensory and motor nerves, which more or less simply find their way from the spinal cord to their peripheral destinations, the nerves of the autonomic system form a complex structure outside of the central nervous system. This consists of two chains of interconnected ganglia alongside the spinal column—a ganglion being a conglomerate of nerve cell bodies, nerve fibers, and connections between them—and another system of more peripheral midline ganglia, the most prominent of which lie on the aorta, the main artery of the body. (Figure 8) All of these ganglia are connected to each other by nerve fiber bundles. From the various ganglia, nerve fibers then go to the organs innervated. The largest of the peripheral ganglia is the celiac ganglion, also known as the solar plexus, which is located under the bottom of the breastbone. This location is a site of sensation in emotional states—though the source of this sensation is not known—and the solar plexus has long been linked to emotion.

The reason for the complex arrangement of the peripheral autonomic ganglia is somewhat obscure. It is clear, however, that this system mediates a variety of complex reflexes on its own. It is influenced by higher control but capable of independent function. The coordinated movements of the intestines that serve to move food along in one direction are an example of this kind of reflex activity. So in this sense the peripheral autonomic nervous system can be said to be a "brain" on its own, like the spinal cord.

What does the autonomic nervous system, whose function seems to be to coordinate basic instinctive functions that keep the body alive, have to do with what we think of as emotion? We usually think of emotion as something

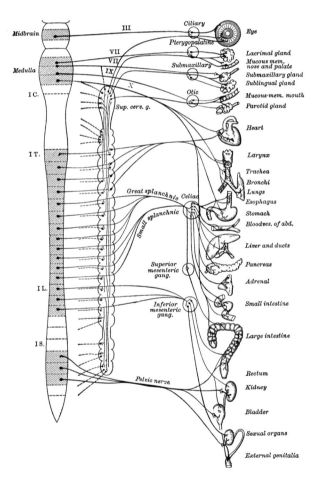

FIGURE 8 • Schematic of the autonomic nervous system

more sublime than the churning of our intestines. Yet the links are clearly there. Our emotional states are often related to instinctive needs—sex being the most obvious example. The state of our digestion has a strong influence on our mood, and vice versa. Our emotional states also manifest through the autonomic nervous system—blushing, goose-pimples, pupillary dilation. Proponents of the idea that we are merely the result of the evolutionary survival of the fittest say that our subjective emotions are simply a byproduct of the instinctive drives that allow us to survive and multiply. The emotional part of the nervous system, including its higher-level portions in the central nervous system, is the coordinator of the body's internal needs, just as the sensory-motor system coordinates its external movements, also subservient to survival and reproduction. Our thinking and even our consciousness, in this view, are also byproducts of this imperative—merely a better way to beat out the competition.

External evidence does little to refute this view. In fact, it is probably correct—partially. It is correct on one level of things. But we have experiences, usually brief, of awe, of wonder, and of meaning: an undeniable feeling-perception that our consciousness itself and its interaction with other consciousnesses are meaningful and part of the scheme of things.

So what then, if we accept our experience of these feelings, is their relationship to our instinctive functions? Gurdjieff's scheme of the three brains, or centers, was put forth in various forms; in one of them[2] he subdivided the lower center into three—instinctive, moving, and sex—and added two higher centers, the "higher emotional" and the "higher thinking." He described these higher centers as being fully formed, but disconnected from their corresponding lower centers. When the functioning of the emotional center reached a certain level of intensity it could "connect" with the higher emotional center. The rare moments of heightened emotional perception mentioned above could be regarded as moments of such a connection. Another way to look at it is that in our usual state of being awake and only reflexively responsive to our environment, but not actually self-aware, our emotional center is not operating at its full capacity, is not fulfilling its potential. Perhaps its vibration rate is too slow. It is operating in an automatic way, more appropriate to the moving center, by means of chains of reflexes.

The autonomic nervous system can also be seen as the regulator of the flow of energy in the body, and this perspective can bring us closer to understanding the relationship between instinctive functions and different kinds of emotion. After we eat, our digestive system is provided with a larger portion of the blood supply, and we may become sleepy; when we run, blood is delivered to the muscles. Our hearts beat faster or slower as required; sugar and other molecular energy-packets are metabolized and distributed according to the needs of the organism. When we perform mental tasks, specific portions of the cerebral cortex that are particularly involved in their execution are given an increased amount of blood. The structures of what I am calling the central part of the emotional brain—between the spinal cord and brainstem and the cerebral cortex—regulate alertness, and are involved in focal attention as well. As discussed above, the neurotransmitters of the emotional system regulate the responses of cortical neurons to other transmitters from more specific circuits, changing the "bias." When I am hungry, food is beautiful. All the diverse aspects of our emotional life can be viewed as connected with the coordination of energy flow. Our perception of the emotional tone of others is of this nature also: they are vibrant or dull; their cheeks are flushed or sallow. Tension or relaxation, the tone of the voice and of the body—these are all manifestations of the configuration of the body's energies, and they are the accompaniments of the subjective emotional state.

Another view is that it is the rhythms of the body and brain that are regulated by the emotional brain: digestion, respiration, heart rate, sleep and waking, and the rhythmic brain activities seen in the electroencephalogram. Rhythms, as discussed earlier, are especially related to the dimension of eternity. They are a necessary backdrop for all activity.

It seems likely that a re-configuration of the body's energies underlies a change in the state of consciousness to one of true self-awareness, in which feelings such as reverence and awe, and the perception of meaning, become possible. This change feels radical: a change of the overall pattern of energy rather than of one or another detail. A change of state in chemistry is a global change—from solid to liquid, for example—and the same is probably true in our organism.

14. How Does a Pattern of Nerve Impulses Become Conscious Perception?

Where along the chain of sensory transformations is it that *I* see? Consciousness remains a mystery to neuroscience, but as discussed in more detail elsewhere,[19] a lower level of perceptual integration, the binding of the different features of an object into a unified percept, seems likely to depend on correlated rhythmic electrical activity between different parts of the brain. For, if there are multiple visual areas specializing in the analysis of color, shape, movement, and so on, how can we have unified percepts? Rather than a *place* in the brain where everything is put together, it seems more likely that a percept is unified in *time*, by correlated rhythmic brain wave patterns between the relevant areas.

A related issue has to do with the location of memory traces. My memory of my grandmother, for instance, includes a picture of her face, the sound of her voice, her name, and a variety of recollections associated with her. Are these all stored together in a cell, or group of nearby cells, somewhere in my brain? More likely the remembering of my grandmother involves the coordinated electrical activity of all the cortical areas that specialize in processing the component features of her memory.

The same diffusely projecting systems from the emotional brain to the cerebral cortex that are involved in changing the "bias" of cortical connections, thus altering our perceptual state, are, not unexpectedly, also involved in the relatively permanent alterations of synaptic strength which underlie memory formation. The long-term alteration of a synapse seems most likely when two things are present: correlated activity between the presynaptic neuron and the post-synaptic one, indicating a significant association between the two; and activation of the post-synaptic neuron by the diffusely projecting systems, corresponding to alertness, attention, or meaningfulness.

15. Self-consciousness

For every moment of self-awareness, there are many of automatic functioning, in which I walk, talk, and carry on my affairs while thinking about this, that, and the other, and experiencing a multitude of transient low-level emotions, mostly negative, without having any real sense of my own existence. These moments leave no real memory. Because of intermittent flashes of greater awareness, which illuminate more than just the moment when they occur, and some knowledge about my habits, I have a sense of continuity. It is this dispersal and disconnectedness of the various inner activities that characterizes my usual semiconscious state. Gurdjieff states:

> Our mind has no critical faculty in itself, no consciousness, nothing. And all the other centers are the same. What then is our consciousness, our memory, our critical faculty? It's very simple. It is when one center specially watches another, when it sees and feels what is going on there and, seeing it, records it all within itself.[20]

So it is a coming together in a particular way of the three brains which, according to Gurdjieff, underlies self-consciousness.

The path to a change in the state of consciousness, to a state in which consciousness truly exists, is, according to all religious teachings, through sustained attention. In different teachings, the idea of sustained attention takes different forms (prayer, concentration, meditation), but the basic discipline is probably the same. An honest attempt to maintain constant attention on any one thing quickly shows how difficult it is and how far from our usual situation. In Christianity, it is described as constant prayer, but not simply an automatic repetition of phrases. In Gurdjieff's teaching, self-remembering plays a central role—the attempt to be aware of oneself at the same time as one is aware of outside impressions. The most accessible approach to this is through maintained awareness of the sensation of one's body. The Buddha said:

> And what is the fourfold establishment of Mindfulness?
>
> The monk, O monks, as respects body, keeps watch upon the body earnestly, intently, clearly conscious, having put away all worldly cares and desires. As respects sensation, he keeps watch upon the sensations, earnestly, intently, clearly conscious, having put away all worldly cares and desires. As respects mind, he keeps watch over the mind, earnestly, intently, clearly conscious, having put away all worldly care and cravings. As respects phenomena, he keeps watch over phenomena, earnestly, intently, clearly conscious, having abandoned all worldly cares and desires.
>
> Again, O monks, the monk, in moving, is aware, "I move"; in standing still, is aware, "I stand still"; in sitting, is aware, "I am seated"; in lying down, is aware, "I am lying down." In whatsoever posture he is placed, he is aware that he is in such and such a posture.[21]

Experience shows that consciousness of this kind, even if sought earnestly, can be measured only in moments, and that even a small increase in the number of moments in which it is present has an extraordinary effect on one's experience of a day. It is reminiscent of the extraordinary amount of energy liberated by the nuclear fusion of a tiny amount of matter: it is a phenomenon from a different level.

Gurdjieff's way of describing the practice of mindfulness is directly related to the idea of three brains: the attention of the mind is brought into relationship with the sensation of the body. Ordinarily, the body carries on with minimal attention, and the mind is occupied with random thoughts and daydreams—they are not connected in the right way. Attempts to maintain attention on sensation can lead, abruptly, to the appearance of self-consciousness, accompanied by a number of phenomena. There is feeling, a sense of meaning. The experience is likely to be vividly remembered. Thought is quieter, and separate from one's sense of self. Usually attention is taken by one thought after another; now there is a separation between this new global attention and thought. One's sense of time changes. There is a sense of being present, and the present has plenty of room in it. Gurdjieff says:

> Consciousness is a state in which a man knows *all at once* everything that he in general knows and in which he can see how little he does know and how many contradictions there are in what he knows.[22]

16. Geometry

One of the intriguing aspects of the nervous system is its geometry. The division of the hemispheres discussed above is an example. The contralateral projections of the cerebral cortex as opposed to the ipsilateral ones of the cerebellum and emotional brain are another. Another interesting fact is that the major sensory modalities of sight, hearing, and touch, are arranged in a somewhat perpendicular manner to each other in the cerebrum. The primary visual cortex is at the back of the head, and the pathways from the eyes travel primarily from front to back. The primary sensory cortex for touch is at the top of the head, and the pathways travel up, through the spinal cord, brainstem, and thalamus. The primary auditory cortex is on the side of the head, in front of and above the ears, and the pathways, coming from the brainstem, travel to the thalamus and then sideways out towards the cortex. Many of the structures of the emotional brain, known as the limbic system, are connected by long loops of fiber bundles that form arcs, arranged in relation to the arcs formed by the fluid-filled ventricles that lie within the brain. (Figure 9)

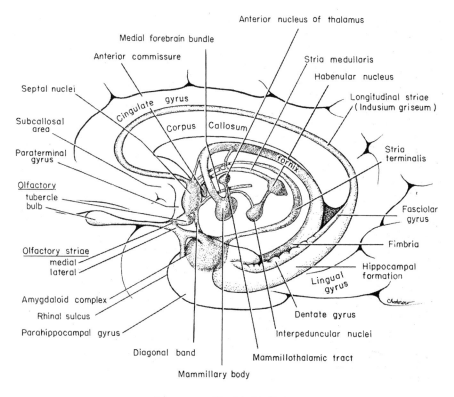

FIGURE 9 • The Limbic System

These arrangements are not thought to have any functional purpose, but are explained as due to patterns of evolutionary and embryological development. The geometry of the body is not thought by modern scientists to have any but a utilitarian purpose—it is simply a convenient way to stack things. It is fairly radical, but not totally absurd, to speculate that in some of these geometrical features may lie clues to aspects of the functioning of the nervous system of which we are now totally unaware. The loops of the emotional brain could suggest an electromagnetic function beyond the mere carrying of nervous impulses. These loops are at right angles to the fibers of the corpus callosum. Until recently, in neuroscientific thinking, the electrical activity of the brain and the body has been regarded as important mainly on a microscopic level. Now, with the idea that correlated rhythmic activity between different parts of the brain is important in perception and awareness, a more global role for electromagnetic brain activity may be emerging. A tiny minority of scientists, such as Robert Becker and Bjorn Nordenstrom,[23, 24] maintain that there are electromagnetic fields on the level of the body as a whole that are important to its functioning. Nordenstrom speculates that some of our responses to weather, for instance, may be due to interactions between these fields and those of the atmosphere. If the body's electromagnetic activity is organized and functional on a macroscopic level, a realm of possible interactions unknown to current scientific thinking may exist.

The formation of global electromagnetic fields of this sort could underlie the change in energy configuration associated with a change in consciousness, and produce a coherence that does not exist otherwise. Meditation produces more coherent brain wave activity[25] and this coherence may even affect other people's brains[26]. Many feel there is a level of interaction among people that is neither verbal nor due to associations set off by sense impressions. The appearance of a moment of greater consciousness has an abrupt aspect to it, suggestive of a sudden change from one vibratory mode to another, although there are variations in intensity and quality. A tuned string vibrates sympathetically when exposed to the note it is tuned to or the note's harmonics. It is perhaps in a similar way that we become sensitive to a different range of perceptions in a state of greater consciousness.

17. Brownian Motion Versus Current Flow

Imagine a container full of particles in Brownian motion, bouncing randomly off the walls and off each other, in which there is an excess of positive charges. The only output of this system is heat. There is an unchanging static electric field around the box due to the excess of positive charge, but the local currents of electricity and their resulting magnetic fields cancel each other out. If more energy is put into the box, there is more heat. Another container near the first contains an excess of negatively charged particles. There is no communication between the two. The second box also produces only heat. Now imagine there is a conductor between the two, with an interposed resistance so that there is not merely a sudden spark discharge. A current flows between the two containers. This current establishes a magnetic field. A new entity, the magnetic field, now exists, even though the two boxes have not appreciably changed. The lines of force of the magnetic field are perpendicular to the electric field—another dimension has been added to the system. This new entity appears and disappears as the connection is made and broken. When it appears, the system is susceptible to magnetic influences; that is, it is capable of receiving information that had no effect on it before. Its perception has changed.

Earlier we asked where the inner world is located, and concluded that its location may be not in space, but in time. A moment of conscious awareness, in which I seem to be aware of the whole of myself, in relation to my surroundings and to the inner activities of my body, feelings, and thoughts, has a sense of unity that we mistakenly conceive of as being the property of a single thing located at a single place. But, as discussed above, the neural components of such an experience do not seem to converge on a single location in the brain; the components of even a perception, or a memory, are distributed throughout much of the cerebral cortex. So, the experiential correlation between consciousness and what we call "presence" is a necessary one; the unity of a conscious experience is a unity in time, not in space— it is what is happening now. The dispersed character of my usual state of waking sleep is a dispersal in time—my thoughts are far away in other times and possibilities, my body dutifully trudges along its linear line of time, my feelings cling desperately now to this thought, now to that impulse or urge. Unity in time is not one-dimensional; if it was, perhaps it would be easier to reach. A conscious moment is not an infinitely thin cross-section of the line of time, for time is not just a line but is three-dimensional. But neither is it a point in three-dimensional time: like a body in space, it must have a certain length, width, and breadth.

If the experience of "I," in the present, consists of a unity in time (now), and depends on a macroscopically coherent electromagnetic pattern, then its generation via simultaneous attention to different sensory modalities (e.g. sensation and vision) can begin to make sense. If a unified percept depends on coherent oscillations between the different brain regions devoted to analysis of its various aspects,[27] then, via the connection of attention, large regions of cortex could oscillate harmoniously. If the sense of "I" is in essence a coherent electromagnetic pattern, then its dimensionality is different from that of our usual state.

Testimonial evidence indicates that Gurdjieff could transfer his "energy" directly to another person. Fritz Peters writes that when he was severely psychologically ill, during the second world war, he went to see Gurdjieff in Paris:

> I remember being slumped over the table, sipping at my coffee, when I began to feel a strange uprising of energy within myself—I stared at him, automatically straightened up, and it was as if a violent, electric blue light emanated from him and entered into me. As this happened, I could feel the tiredness drain out of me, but at the same moment his body slumped and his face turned gray as if it was being drained of life. I looked at him, amazed, and when he saw me sitting erect, smiling and full of energy, he said quickly: "You all right now—watch food on stove—I must go." There was something very urgent in his voice and I leaped to my feet to help him but he waved me away and limped slowly out of the room.
>
> He was gone for perhaps fifteen minutes while I watched the food, feeling blank and amazed because I had never felt any better in my life. I was convinced then—and am now—that he knew how to transmit energy from himself to others; I was also convinced that it could only be done at great cost to himself.
>
> It also became obvious within the next few minutes that he knew how to renew his own energy quickly, for I was equally amazed when he returned to the kitchen to see the change in him; he looked like a young man again, alert, smiling, sly and full of good spirits. He said that this was a very fortunate meeting, and that while I had forced him to make an almost impossible effort, it had been—as I had witnessed—a very good thing for both of us. He then announced that we would have lunch together—alone—and that I would have to drink a "real man's share" of fine old Armagnac.[28]

18. Summary

Matter, energy, and concept (idea, form, information) make a basic triad that applies to all things. These three aspects of everything can be related to three dimensions of time: linear time, eternity, and possibility. Our three brains, serving movement, emotion, and thought, seem to be specialized along these lines. True consciousness requires a special interaction between them. This interaction produces a change in the dimensionality of our inner lives. Only the moving brain can be thought of as fully developed, as we are. Automatic movements are appropriate, but automatic thoughts are a shadowy one-dimensional projection of the full capability of thought. The special interaction between the brains realized in states of greater consciousness changes their level of functioning. The state of self-awareness I have been describing seems to relate to the proper functioning of the emotional brain in which its "higher emotional" aspects are included, which may correlate with a change of the electromagnetic patterns of the brain and body to one of macroscopic coherence. Gurdjieff describes a further state of consciousness, "cosmic consciousness," which is the proper consciousness of the "higher thinking" center. I cannot even speculate on what further interactions take place to bring this about, but the picture, in outline and very tentatively, begins to make sense.

> Time is a three-fold stream passing through our three centers (mind, emotions, instinct). We fish in that stream. What we catch is ours. What we don't catch is gone.[29]

I cannot tell whether the idea of multiple time dimensions as proposed by Ouspensky and discussed here will prove fundamental, or turn out to be merely a metaphor. Similarly, whether the idea of three brains is of basic importance, or is an arbitrary division, remains to be seen. The ideas put forth here, while grounded in both inner and outer facts, are far away from current scientific understanding. They do not constitute a theory; rather they form a speculation, with many loose ends. But the issues addressed are fundamental and require confronting. Many others have confronted and are confronting them; understanding is elusive. But our task as humans, more than anything else, seems to be to attempt to understand. Perhaps our consciousness is like a lens through which the universe is focused, in a return movement towards the source of all things.

Notes

1. G.I. Gurdjieff, *All and Everything, First Series, Beelzebub's Tales to his Grandson* (New York, E.P. Dutton & Co., 1964).

2. P.D. Ouspensky, *In Search of the Miraculous: Fragments of an Unknown Teaching* (New York, Harcourt, Brace & World, Inc., 1949).

3. P.D. MacLean, *The Triune Brain in Evolution: Role in Paleocerebral Functions* (New York, Plenum Press, 1990).

4. G. Montgomery, "The mind in motion." *Discover*, March 1989, pp. 58–68.

5. Gurdjieff, *op cit.*, p. 124.

6. P.D. Ouspensky, *A New Model of the Universe* (New York, Alfred A. Knopf, 1934), p. 119.

7. J.B. Priestley, *Man and Time* (London, Aldus Books, 1964).

8. J.T. Fraser, *Time: The Familiar Stranger* (Amherst, MA, University of Massachusetts Press, 1987).

9. Ouspensky, *In Search of the Miraculous*, p. 213.

10. See "The Light of the Beholder" in this volume.

11. C.J. Isham, personal communication.

12. S. Hawking, *The Universe in a Nutshell* (New York, Bantam Books, 2001).

13. Ouspensky, *In Search of the Miraculous,* p. 107–108.

14. Kandel ER, Schwartz JH (eds.), *Principles of Neural Science*, 2nd ed. (New York, Elsevier, 1985), p. 521.

15. See "Shadows of the Real World" in this volume.

16. G.M. Edelman, *The Remembered Present: A Biological Theory of Consciousness* (New York, Basic Books, Inc., 1989).

17. N.D. Cook, "Callosal inhibition: the key to the brain code." Behav. Sci. 29:98-110, 1984.

18. S.J. Dimond, "Depletion of attentional capacity after total commissurotomy in man." Brain 99:347-356,1976.

19. See "Shadows of the Real World" in this volume.

20. G.I. Gurdjieff, *Views from the Real World* (New York, E.P Dutton & Co., Inc., 1973), p. 271.

21. "Setting-Up of Mindfulness," *The Bible of the World* (New York, Avon Books, 1973), p.244.

22. Ouspensky, *In Search of the Miraculous*, p. 155.

23. R.O. Becker and G. Selden, *The Body Electric: Electromagnetism and the Foundation of Life* (New York, William Morrow, 1985).

24. B.E.W. Nordenstrom, *Biologically Closed Electric Circuits* (Stockholm, Nordic Medical Publications, 1983).

25. M. Murphy, S. Donovan, *The Physical and Psychological Effects of Meditation, A Review of Contemporary Meditation Research with a Comprehensive Bibliography, 1931–1988* (San Rafael CA, Esalen Institute Study of Exceptional Functioning, 1988).

26. D. Orme-Johnson, M. Dillbeck, R.K. Wallace, "Intersubject EEG coherence: is consciousness a field?" Intern. J. Neurosci. 16:203-209, 1982.

27. W. Singer, "Self-organization of cognitive structures." In J.C. Eccles, J. Creutzfeldt (eds.), *The Principles of Design and Operation of the Brain* (New York, Springer-Verlag, 1990), pp. 119–128.

28. F. Peters, *Gurdjieff Remembered*, (New York, Samuel Weiser, 1974), p.82–83.

29. A. R. Orage, quoted by Louise Welch, in *Orage with Gurdjieff in America* (Boston, London, Melbourne and Henley, Routledge & Kegan Paul, 1982), pp. 79–80.

THE COSMIC METABOLISM OF FORM

It is fairly obvious by now that life on earth forms a vast interconnected and interdependent network. Organic molecules are recycled over and over, those of one organism providing food for another. Bacteria transform our waste into soil. Carbon dioxide is taken in by plants and oxygen released; in animals the reverse process occurs, in an endless cycle of energy exchange, ultimately fueled by the sun.

Ecosystems depend on a delicate balance of mutually supportive interactions, forming a greater organism. For that matter, any given organism is itself an ecosystem: birds keep the hippopotamus clean, our intestinal bacteria help us digest. Even the subcellular organelles that power our cells, the mitochondria, were, it is thought, once independent organisms. The boundaries of selfhood become blurred and somewhat arbitrary.

These interrelations extend out to the entire universe. The sun, besides providing the energy that fuels life on earth and keeping the solar system in place with its gravity, also interacts with the earth and other planets through the solar wind, shaping the earth's magnetic body and influencing the weather, which in turn affects all earth's creatures. There are certainly other crucial interactions at this electromagnetic level; the more we look, the more we find a complex and delicate order. And the basic elemental building blocks of our world and of organic life—carbon, oxygen, nitrogen and the rest—were formed in the nuclear furnaces of stars and distributed by the explosions of supernovae, as part of vast cosmic cycles of stellar formation, growth, and death.

In considering these things, we generally think in terms of substances being endlessly recycled and transformed, in large and small metabolic cycles. But the consequence of all these transformations is to preserve form. An organism, or an ecosystem, is a form, whose elements are constantly changing while its structure is preserved for a time: its lifetime. It is estimated that every single atom in the human body is replaced in at most seven years. Most cells, with the exception of many of the cells of the nervous system, are also replaced during the lifetime of the body, some more rapidly, like skin and intestinal cells, and others more slowly. Similarly, the organisms in an ecosystem live and die, to be replaced by others, but the ecosystem has a longer lifespan.

A subtler, often unacknowledged, materialistic bias in our thinking appears in the assumption that form is simply the result of the properties of matter. Atoms form molecules, molecules cells, cells organisms, etc., because the basic properties of the smaller constituents determine how they organize themselves into the larger assemblies. Given the physical laws that govern the fundamental particles and forces, and time for evolutionary processes to work, everything has become what it is now, culminating in self-aware organisms that happen to be able to contemplate these matters. It is certainly true that quarks and the strong nuclear force must have the properties they have in order to form nuclei, and that oxygen, carbon, nitrogen, and hydrogen can join in the molecules they form only because of their specific chemical properties, and so on at every level. But fundamentally, what is matter? Physics has long since reached an impasse in defining it with any kind of bricks and mortar, or even whirling particles, picture. Atoms are mostly empty space, and the subatomic constituents of everything are both waves and particles, but not both at the same time, or perhaps neither, but with properties of both. Ultimately, the only reliable descriptors of the constituents of matter are mathematical equations, the abstract mathematical forms of group theory. So form underlies substance, as Plato taught, rather than the reverse.

If this is so, perhaps intelligent beings such as ourselves, who are capable of resonating with (and therefore discovering) the mathematical forms that govern external reality, do not have these capacities merely as an accidental byproduct of being the fittest organisms in the evolutionary struggle, but have a more fundamental role in the universe. Just as our bodies transform substances in metabolic cycles of varying complexity, and take part in the larger ecological metabolic cycles of organic life on earth, our minds take in and manipulate, transform, break down and build up, in short metabolize, forms—as impressions, perceptions, concepts, memories, and plans. And we take part in a larger ecology of form: culture, literature, art, science.

One could say that everything in the universe has three aspects: a physical or material aspect, an energetic aspect, and a form. We share with other creatures the ability to participate in the metabolism of matter and energy, but seem uniquely endowed by our consciousness to participate in the metabolism of form.

In the view of the traditional religious teachings, human beings have a special place in the cosmos:

> And God said, let us make men in our own image, after our likeness: and let them have dominion over the fish of the sea, and over the fowl of the air, and over the cattle, and over all the earth, and over every creeping thing that creepeth upon the earth. . . . And God said unto them, be fruitful, and multiply, and replenish the earth, and subdue it. (Genesis 1:26-28)

One of the main concerns of the ecologically minded is that humanity is despoiling the earth and disrupting the web of life. But even this capacity implies its converse: "to replenish the earth." This is not simply to populate it, but to care for it and maintain it, by virtue of our understanding, in ways that are obvious and perhaps in other ways that we now only dimly perceive.

Human intelligence provides us with a unique relationship to time: we are not confined to the relentless march of our bodies through linear time, but are able to remember the past, foresee the future, and contemplate all of their possibilities. We can look at a tree and see planks and beams and a bridge, look at mud and see adobe bricks and a house. We can create, repair, and maintain to a degree and with a flexibility unknown to other creatures. We are, at our best, anti-entropy machines.

This capacity of our consciousness is evident in all our artifacts, and whether ultimately it will serve to replenish the earth or to destroy it is still in the balance. But it is possible that our awareness has a purpose on another level as well. Some religious and mystical teachings hold that human consciousness, in reflecting reality, also helps to bring it into being. According to the Islamic mystic Ibn Arabi, man acts "as the eye through which God can see His own creation":[1]

> He praises me (by manifesting my perfections and creating me in His form),
> And I praise Him (by manifesting His perfections and obeying Him).
> How can He be independent when I help and aid Him? (because the Divine attributes derive the possibility of manifestation from their human correlates).
> For that cause God brought me into existence.
> And I know Him and bring Him into existence (in my knowledge and contemplation of Him)."[2]

A similar concept, though with a less hierarchical emphasis, appears in Hindu teaching as Indra's net:

> In the Heaven of Indra, there is said to be a network of pearls, so arranged that if you look at one you see all the others reflected in it. In the same way, each object in the world is not merely itself but involves every other object and in fact IS everything else.[3]

Perhaps the most detailed, though still incomplete, elaboration of this idea is contained in the writings of G.I. Gurdjieff. According to Gurdjieff, humans take in three kinds, or levels, of food: ordinary food, air, and impressions. These consist of substances of increasing fineness, but each is metabolized according to the same general laws. The three foods and their metabolites interact with each other, helping each other's digestion: the clearest example is the necessity of oxygen for the complete metabolism of ordinary food, which is well known to ordinary physiology and biochemistry. On first encountering this idea, the analogy between the metabolism of ordinary food and the taking in and processing of impressions by the brain may seem far-fetched, but there are clear parallels. The food we eat consists mostly of macromolecules—proteins, carbohydrates, and fats—which themselves are compounds of smaller molecules: amino acids, sugars, and fatty acids. These in turn are made up of different configurations of atoms, mainly carbon, oxygen, nitrogen, and hydrogen. The process of digestion is a progressive breakdown of these large macromolecules, which contain energy stored in their atomic arrangements, into much smaller molecules, primarily water and carbon dioxide, the atomic arrangements of which are in a much lower energy state, thus releasing the energy of food for use by the body. Some of the molecular constituents of food are also recycled by the body and built up again into the proteins, carbohydrates, and fats of the body itself. So the metabolism of food has at least two results: one on the same level, which is the maintenance of the body by producing substances, just like those in food, which make up our flesh; and one on a "higher" level, which is the production of energy to maintain the life of the body.

Impressions also consist of larger configurations of smaller elements. A visual scene is made up of many shapes and colors, which in turn are made up of lines and boundaries between patches of differing contrast and hue. Colors themselves can be thought of as consisting of mixtures of primary colors and different amounts of white, gray, and black, and ultimately different combinations of wavelengths of light. Complex sounds are combinations of fundamental frequencies varying over time. In processing impressions, the brain breaks them down into more fundamental components. In the visual cortex, individual cells are activated by the elementary components of the visual scene: lines and edges, spatial frequencies, binocular disparity conveying depth, basic colors. The analogy is clear: impressions are also metabolized via a process of breakdown into smaller and simpler elements.

Subsequently, impressions are, so to speak, reconstituted, into our perceptions. Something has been added: our awareness and our interpretations, based on experience. The capacity to perceive depends on a long process of learning in early childhood. Our brains are partly formed, or tuned, by our environment of impressions. People born blind whose sight is restored later in life typically do not learn to see properly, although the eye is transmitting all the information to the brain; at best they are like people using a dictionary to speak in a foreign language, painfully correlating their visual impressions with their learned perceptions in other sensory modalities.

The reconstituted perceptions are internal representations of the external stimuli, analogous to the reconstituted macromolecules that make up our bodies. What do these internal representations consist of, in material terms? This is an as yet unresolved question scientifically, but evidence suggests that their physical correlates may be complex patterns of electromagnetic vibrations in the brain. Visual configurations are conveyed to us via patterns of electromagnetic vibrations, light, and our internal perceptions may be of the same materiality.

According to Gurdjieff, our impressions are not fully metabolized in our ordinary state. For their metabolism to proceed further, we must be conscious of ourselves in the act of receiving impressions, just as oxygen is necessary for the full metabolism of food. The full metabolism of food releases much more energy than that released in the absence of oxygen (anaerobic metabolism). What kind of energy is released by the full metabolism of impressions? Perhaps the energy of consciousness. Although consciousness is required for this process to take place, acting like a catalyst, the result is the production of even more consciousness. Or more precisely, it is the addition of consciousness to the impressions, just as the energy of life is added to the macromolecules of our bodies.

This vivification of impressions feeds our inner life, which needs conscious impressions to grow, and may also serve a larger purpose, enabling God to "see" his own creation through us and other conscious observers throughout the universe.

The great paradox of quantum theory is that the form of what we observe in looking at the smallest constituents of matter, for instance whether they consist of waves or particles, depends on the kind of observation that is conducted. Some argue that the "decision" is made when a conscious observer appears, others that a measuring apparatus is sufficient, but in any case, the measuring apparatus was made by a human being.[4] Perceptions have a similar property, for when we see something we are making decisions. This is made evident in the perception of ambiguous figures such as the Necker cube (Figure 1), but is going on all the time.

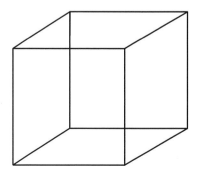

FIGURE 1 • The Necker cube can be seen in two ways, with either the lower face forward, or the upper face. Each way results in a distinct three-dimensional interpretation of the cube. One can easily switch from one perception to the other, but not see both at once, or easily fail to see either. This is a simple example of the difference between the external pattern that falls on the eyes and the resulting perception, involving a perceptual decision.

So perhaps, in an as yet unclear fashion, we are, or can be, part of a great cosmic ecology of consciousness, maintaining the form of the universe against the entropy of linear time because of our potential capacity for metabolizing form, which is made possible by our flexible and encompassing relationship to time itself. In this process, we become part of everything on a conscious level, just as we are part of everything on the level of gross materiality. Just as our bodies are made up of atoms that once were created in supernovae, and passed through a variety of inorganic and organic entities, the material of our inner lives consists of all our perceptions—of other beings, of earth, sea, sky, and stars, and of the fundamental laws that govern it all—and we are thereby connected to everything on both levels, and to the Whole.

> "He through Whom we see, taste, smell, feel, hear, enjoy, know everything, He is that Self.
>
> "Knowing That by Which one perceives both dream states and waking states, the great, omnipresent Self, the wise man goes beyond sorrow.
>
> "Knowing that the individual Self, eater of the fruit of action, is the universal Self, maker of past and future, he knows he has nothing to fear.
>
> "Born in the beginning from meditation, born from the waters, having entered the secret place of the heart, He looks forth through beings. That is Self.
>
> "That boundless power, source of every power, manifesting itself as life, entering every heart, born with the beings, that is Self."[5]

Notes

1. R. Landau, *The Philosophy of Ibn 'Arabi* (London, George Allen & Unwin, Ltd., 1959), p. 74.

2. Ibn 'Arabi, with commentary in parentheses by 'Abdu 'l-Razzaq al-Kashani, quoted in ref. 2, p. 74.

3. Charles Eliot, quoted at www.cs.kent.ac.uk/people/staff/saf/networks/networking-networkers/indras-net.html.

4. There are other theories, and the question of how the wavefunction "collapses" is not really resolved.

5. *Katha Upanishad*, edited by the author based on multiple translations.